OUR ACOUSTIC ENVIRONMENT
   Frederick A. White

ENVIRONMENTAL DATA HANDLING
   George B. Heaslip

THE MEASUREMENT OF AIRBORNE PARTICLES
   Richard D. Cadle

ANALYSIS OF AIR POLLUTANTS
   Peter O. Warner

ENVIRONMENTAL INDICES
   Herbert Inhaber

ENVIRONMENTAL INDICES

ENVIRONMENTAL INDICES

HERBERT INHABER

ENVIRONMENT CANADA
OTTAWA, CANADA

A WILEY-INTERSCIENCE PUBLICATION

JOHN WILEY & SONS

NEW YORK • LONDON • SYDNEY • TORONTO

Library of Congress Cataloging in Publication Data:

Inhaber, Herbert, 1941–
  Environmental indices.

  (Environmental science and technology)
  "A Wiley-Interscience publication."
  Bibliography: p. 164–166
  Includes index.
  1. Environmental indexes. I. Title.

GF23.I53153   301.31   75-34290
ISBN 0-471-42796-9

Printed in the United States of America

10 9 8 7 6 5 4 3 2 1

For my wife, Elizabeth

SERIES PREFACE
Environmental Science and Technology

The Environmental Science and Technology Series of Monographs, Textbooks, and Advances is devoted to the study of the quality of the environment and to the technology of its conservation. Environmental science therefore relates to the chemical, physical, and biological changes in the environment through contamination or modification, to the physical nature and biological behavior of air, water, soil, food, and waste as they are affected by man's agricultural, industrial, and social activities, and to the application of science and technology to the control and improvement of environmental quality.

The deterioration of environmental quality, which began when man first collected into villages and utilized fire, has existed as a serious problem under the ever-increasing impacts of exponentially increasing population and of industrializing society Environmental contamination of air, water, soil, and food has become a threat to the continued existence of many plant and animal communities of the ecosystem and may ultimately threaten the very survival of the human race.

It seems clear that if we are to preserve for future generations some semblance of the biological order of the world of the past and hope to improve on the deteriorating standards of urban public health, environmental science and technology must quickly come to play a dominant role in designing our social and industrial structure for tomorrow. Scientifically rigorous criteria of environmental quality must be developed. Based in part on these criteria, realistic standards must be established and our technological progress must be tailored to meet them. It is obvious that civilization will continue to require increasing amounts of fuel, transportation, industrial chemicals, fertilizers, pesticides, and countless other products and that it will continue to produce waste products of all descriptions. What is urgently needed is a total systems approach to modern civilization through which the pooled talents of scientists and engineers, in cooperation with social scientists and the medical profession, can be focused on the development of order and equilibrium to the presently disparate segments of the human environment. Most of the skills and tools that are needed are already in existence. Surely a technology that has created such manifold environmental problems is also capable of solving them. It is our hope that this Series in Environmental Sciences and Technology will not only serve to make this challenge more explicit to the established professional but that it also will help to stimulate the student toward the career opportunities in this vital area.

Robert L. Metcalf
James N. Pitts, Jr.
Werner Stumm

PREFACE

"Everyone talks about the environment, but nobody does anything about it." If this statement is true, it's because midway between talk and action must come understanding. This understanding is lacking in most discussion of today's environment. We're told by some groups that lungs, leaves, and lakes have never had it so good. Others say that our lungs are black with dirt, the leaves brown with imminent death, and the lakes green with algae. Both sides can't be right, and many of us wonder where the truth lies.

Environmental indices are a way of finding the truth. Simple mathematical tools, they help us understand how the state of the environment compares to desirable conditions. But don't let the mathematics throw you. What's important is not the particular equation used, but the judgment reached. Environmental indices can quiet down the harangues of both sides of an environmental issue.

Indices are used for a wide variety of environmental topics. The chapters are arranged so that those interested in particular topics can find them easily. For example, there are chapters on air, water, and land, as well as on such less well-studied aspects as vibration and odor. Although the subject of environmental indices is relatively new, it has been or can be applied to many ecological fields.

The number of axes that have been ground over environmental issues are enough to clear-cut every tree on earth. Out of the dust and commotion of ax-grinding has emerged little light, as far as the public is concerned. This book was written to show a better way of understanding the condition of our environment. Where there is understanding, there can be real improvement— which is what we all want.

The playwright Christopher Fry once cried out for longitudes where there are no platitudes. Let's use environmental indices to eliminate mis-, half-, and other untruths about our surroundings.

HERBERT INHABER

New Haven
October 1975

CONTENTS

1. **Introduction**   1

   A Day in the Life of an Index   1
   How an Index Works   1
   Do We Really Need New Indices?   2
   Environmental Data and Standards—How Much Is Enough?   5
   How Much Is Background?   7
   Environmental Indices Aren't Just Pollution Indices   8
   What About Energy Indices?   9
   Combining Indices   10
   A National or International Index   12
   Not Just Numbers, but Action   13

2. **Economic Indices**   14

   The Way Economic Indices Came About   15
   The Dollar—The Almighty Standard   19
   Two Types of Economic Index   20
   Weighting of Indices   23
   Quantity and Quality   27
   Nobody Is Average to an Index   28
   Is an Environmental Index More Difficult to Produce?   28
   They Aren't So Perfect after All   29

3. **What's Been Done So Far?**   31

   The First Environmental Index—Health   31
   All the Indices Fit to Print—The Newspapers   33
   The National Wildlife Federation Index   34
   AAAS Symposium   35
   The Role of Independent Groups—MITRE and Others   36

xi

Does a Government Dare?   39
International Indices   45

4. **Air Quality Indices**   **46**

Why Air Indices Came First   46
What Am I Breathing, Anyway?   48
How Standard Is a Standard?   49
Changing Standards   52
Linear or Curved?   53
Scales—Up, Down, and Sideways   54
Synergy   55
Two Approaches to Air Quality Indices   57
Measured Air Quality Indices   58
Estimated Emission Indices   60
Can the Two Types of Index Be Combined?   61
Seeing Is Believing   62
A Smell by Any Other Name   63
Achtung! Achtung!   64
Other Types of Air Indices   64
Some Typical Air Indices   64
Conclusions   66

5. **Water Indices**   **67**

Different Uses—Different Abuses   67
A Solution to Pollution Is Dilution   69
To Prepare, Just Add Water—Parameters and Pollutants   70
Measurements and Emissions   71
The River Basin—The Natural Geographic Unit   72
Water Quality Indices in France   74
Canadian Water Quality Index   75
Manchester Water Quality Indices   77
Water Indices in Japan   78
American Water Quality Indices   80
Conclusions   83

6. **Land Indices**   **84**

While Strolling in the Park One Day . . .   84
Having a Wonderful Time—Recreation Indices   87

When You Can't See the Woods for the Cut-Down Trees— Forestry Indices 89

Standing Room Only—Population Densities 92

Quality of Soils and Erosion 94

Dig We Must?—Indices of Strip-Mining 98

Down in the Dumps 99

'Tis a Far, Far Better Thing—Taking Account of Distances 100

Conclusions 102

7. Biological Indices **103**

The Contrariness of Living Creatures—Measuring Things That Run Away 103

Biological Sensitivity 105

I'll Take One of That, Three of Those—Biological Diversity 108

Where the Deer and the Antelope Don't Roam—Wildlife Indices 111

Little Fishes in the Brook 114

Death in the Afternoon—and Morning and Night 117

Lichens—Like 'Em or Not 118

Other Biological Indices 119

Conclusions 120

8. Aesthetic Indices **121**

At Last, a Use for Critics 121

His Yard is Messy, Mine is Only Lived In 123

How Sweet it Isn't—Indices of Odor 125

An Index We Can't Refuse—Garbage and Other Litter 129

Natural and Man-Made Beauty 131

Conclusions 133

9. Other Environmental Indices **134**

Rays from Alpha to Gamma—Radioactivity Indices 134

You Know I Can't Hear You—Noise Indices 139

Shake, Rattle, and Roll—Vibration Indices 143

Social and Environmental Indices 144

On-the-Job Health 145

Beyond the GNP—Social Indicators 146

Two Can Play that Game—Indices and Impact Statements 148

Conclusions 153

10. **Where Do We Go**
     From Here?   **154**

     How the General Public Fits In   154
     Trial and Error, but Not Too Much of the Latter   158
     I See a Tall, Dark, Handsome Stranger—Prediction of Indices   159
     Getting Your Environmental Dollar's Worth   161
     Environmental Motion from a Notion   162

Bibliography   **164**
Index   **167**

1

INTRODUCTION

A DAY IN THE LIFE OF AN INDEX

You roll out of bed in the morning, and as you're sleepily preparing your coffee, you vaguely hear a voice over the radio giving the day's predicted air quality indices. A few hours earlier down the block, a group of hopeful fishermen was poring over yesterday's newspaper for the water quality indices at their favorite lakes. Buried elsewhere in the paper is a stuffy announcement from the city fathers claiming that their proposed new park would boost the recreational index for the town by 10%. A few hours from now, high government officials in the capital will gather to discuss the latest month's overall environmental indices for the region and the nation. Later in the evening, a conservation group will meet to deplore the levels of many environmental indices and to urge stronger action to improve them.

Sound far-fetched? As of the moment, it is. There are very few people who now know what an environmental index is, let alone act on it, but within the next few years it's likely that these indices will be used as the basis for action by governments, antipollution groups, industry, and the public. The object of this book is to tell you what they are, the way they work, and how they can be used to improve our environment. We need indices not to while away long winter nights scanning rows of figures, but because we seek relief from environmental degradation.

HOW AN INDEX WORKS

Before we can discuss the details of environmental indices, we have to know what we're talking about. The term *environment* is one that most of us intuitively feel we understand, but what is an *index*?

1

The word may conjure up visions of back room economists playing with slide rules, but the concept itself is really quite simple. In fact, we use indices of one sort or another every day without thinking about them. Dictionaries contain complicated definitions of the word, but a simplified one might be "the comparison of a quantity to a scientific or arbitrary standard." How do *you* use indices? Consider the thermometer that you peer at each day. The two scientifically fixed points on it, 32° for freezing and 212° for boiling of water, are the standards. The actual numbers are arbitrary. They might just as well be 0° and 100°, as they are in the centigrade system. The room temperature is somewhere between 32° and 212° if the furnace isn't acting up again. To find the numerical value of the temperature, you compare the quantity (the length of the red line above 32°) to the standard (the distance between 32° and 212°). Viola! You've been using an index without knowing it.

Of course, economic measures like consumer price indices get more publicity than thermometers, but they're calculated along the same general lines. We'll discuss them in the next chapter, but a few words of description are worthwhile here.

To calculate a price index, a scale is devised. To illustrate this concept, let's consider the mythical land of Gourmetia, where the population lives solely on truffles and champagne. Suppose that the average weekly food bill is 60 crowns for truffles and 40 crowns for champagne. Then inflation hits Gourmetia. The cost of the typical weekly basket of truffles goes up to 70 crowns, and the bill for washing them down goes to an unprecedented 60 crowns. What is the food price index?

We compare the new price to the old, that is, the quantity in question to the base (or standard). The latter is the "usual" bill of 60 + 40 = 100 crowns. The new bill is 70 + 60 = 130 crowns, so the index is then 130/100 = 1.30, or a 30% increase. Figure 1 shows the calculation.

Our indices, both economic and environmental, are more complicated than those for Gourmetia. We buy everything from hot dogs to hair dryers, and this is reflected in our price indices. However, the concept remains the same: the comparison of a quantity to a prespecified base.

DO WE REALLY NEED NEW INDICES?

With all the indices being bandied about, from stock market indices to the humidex (a measure of the combined effect of heat and humidity),

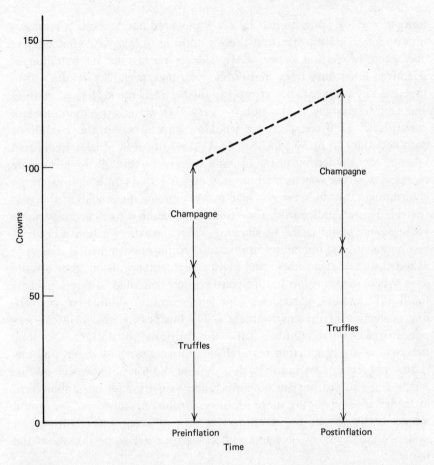

FIGURE 1
*Calculation of a price index for the residents of Gourmetia. The base
to which subsequent prices are compared is that of preinflation days.
At that time, only 100 crowns per day were needed for an adequate
diet of champagne and truffles. In the postinflation period 130 crowns
were required. Thus, the value of the index in the bad new days is
130/100, or 1.3.*

you might think that another set of indices, this time dealing with the
environment, would be as useful as adding another floor to the Empire
State Building. Are environmental indices necessary?

They are—if we really want to know the state of our environment.
All too often we are bombarded with conflicting information on pollution problems. A conservation group claims that in a particular area of

concern, we are going to hell in a V-8 powered handbasket. A corporation or municipality says that there's nothing to fear, and that matters may even be getting better. Each side is backed by its battalion of scientists, laboratory tests, and public relations specialists. In the war to save the environment, the group caught between the volleys of charges and countercharges is the public. Very few of us are environmental scientists. Even if we were, we wouldn't have access to the specialized data necessary to draw conclusions. As a result, most of us are confused when a new battlefront opens up, and even more befuddled when trying to assess how the war as a whole is going. A set of simple indices on the environment would serve to clear away some of the smoke of the battlefield. If used judiciously, they could also be used to clear away some of the smoke of our cities, by showing when and where action is needed.

There's at least one more significant need for environmental indices—as a measure of government and private performance. Long gone are the days when spending on nature conservation consisted of buying shorts for Boy Scouts and barbed wire for fences around wildlife refuges. Today protection of the environment is big business, with worldwide expenditures easily running into the billions of dollars. Nations everywhere are conducting research, monitoring air and water, and imposing regulations on industry that require the use of expensive instruments and equipment. But how much are we getting for our dollar, yen, or franc? Does spending more on the environment improve it, or would we get better results by spending less in a different way? Only by an understandable set of objective indices can we get some inkling of the truth.

We already use economic indices to measure important aspects of government performance. The monthly unemployment and consumer price indices are seized upon as a measure of how well our public financial planners are steering between Scylla and Charybdis. Stock market and capital investment indices are used as a gauge of how well the private sector is doing. The use of environmental indices will be more difficult, for reasons we'll discuss in the following chapters, but their use should lift a little of the verbal fog we're immersed in now.

How would these new indices work? As an example, suppose that the index of oxides of filthium, a hypothetical part of the air quality index, rose over a period of a few months from 0.5 to 0.6. This would indicate a significant increase of these oxides in the air. It would then be up to government, industry, and the public to agree on what measures were needed to decrease the value of the index. These could include better methods of removing oxides of filthium from car exhausts, power plant stacks, or filthium-burning motors. The index itself doesn't give us the final answer to a problem, any more than the broken glass of a fire alarm box after the signal is sounded can put out the flames. But it can tell us where the trouble is.

ENVIRONMENTAL DATA AND STANDARDS—HOW MUCH IS ENOUGH?

An index is essentially a fraction, and to produce one, we need a numerator and a denominator. The numerator is the measurement of the quantity we're interested in, and the denominator is the standard we're comparing it to. In devising an index, like the song about love and marriage, you "can't have one without the other." Figure 2 shows graphically how they can vary. Do we have enough data to produce these fractions?

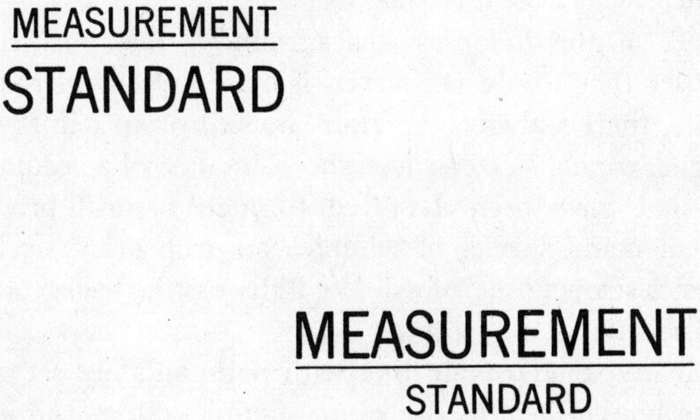

<u>MEASUREMENT</u>
STANDARD

<u>MEASUREMENT</u>
STANDARD

FIGURE 2
Any index is essentially a fraction. The quantity to be measured is the denominator, and the standard—whether it's the number of people in the labor force or the allowable concentration of sulfur dioxide in the air—is the numerator. When the measurement is much less than the standard, as in the upper part of the figure, the value of the index is low and matters are usually fine. When the measurement is larger than the standard, as in the lower part, the index value is high, which indicates problems.

Surprisingly enough, the answer is not a resounding yes. Gathering data is an expensive process. To obtain a perfect measure of air quality, we'd need instruments at every street corner. However, the cost would make taxpayers hot under the collar even during February blizzards. In practice, we have to settle for a few air quality instruments in each city. They should be located in industrial, commercial, and residential areas, giving a city-wide average of air quality.

It might seem highly inaccurate not to have pollution measuring devices everywhere, but nature usually averages itself out enough so that this isn't really necessary. For example, if you wanted to find the temperature inside your home, you probably wouldn't hang thermometers on every wall. One, two, or possibly even three would give you a fairly good idea of the temperature distribution. It's true that pollution levels tend to vary more dramatically than temperatures, but instruments are often placed where the more extreme values are likely to be, such as near factories, freeways, and the like. The following chapters discuss in more detail the limitations of the available data.

The problem of environmental standards is considerably more complex than that of data. We all know that too much dust in the air is bad, but how much is allowable? What should be the standard? The problem is complicated by at least three factors.

First, part of the index is "background"; this part results from natural causes that would exist even if man had never trod the earth. For example, there's always a certain amount of dust in the air due to volcanoes and winds. Water sometimes contains what could be termed pollutants that have been dissolved through natural processes. The population of many species of wildlife can drop dramatically through bad weather destroying their food. We'll discuss the background effect in more detail shortly.

Second, many standards dealing with pollutants are set on the basis of danger to human health. But quite often it's difficult to judge where to draw the line. Let's take an example. Suppose that a concentration of the hypothetical filthium oxides of 1.0 parts per million (ppm) of air is medically shown to cause sore eyes in 1 person in 10. Obviously, a safe standard would have to be considerably lower than that. Down we go to a concentration of 0.5 ppm, which causes eye problems in 1 person in 100. At a concentration of 0.3 ppm 1 person in 10,000 is irritated. Somewhere along the line a trade-off has to be made between allowing a few highly sensitive people to have problems and obtaining an enforceable and reasonable standard. Since this is obviously a difficult decision, governments often have been reluctant to set these standards.

Third, by the very nature of some aspects of the environment, standards are difficult to construct. For example, we all agree that litter lowers the quality of the environment. In spite of our efforts at recycling, the cylindrical aluminum flower is often the most common variety by our roadsides. But how can we construct a standard for litter? Would it be the number of cans per mile of road, or the area of paper wrappers per square yard? Nobody has simple answers to these and similar questions, although they are needed. We return to this question in Chapter 8.

We don't yet have enough data and standards to construct indices for every aspect of the environment, although we've made progress in recent years. This short discussion should illustrate why the road can be rocky.

HOW MUCH IS BACKGROUND?

As we mentioned before, part of the level of many environmental indices is background, due to natural causes. Unfortunately, we can't give a general rule for the fraction which is natural and that which is manmade, since it varies so much from index to index.

To choose an analogy from the economic realm, this background effect is as if tiny gremlins crept in at night to the stores across the land and increased the price of everything we buy, and their changes couldn't be rectified in the morning. This "gremlin background" would affect the consumer price index. We couldn't separate the gremlin-produced changes from the normal ones unless we noted the prices before the stores closed.

It's very easy to say that further research will allow us to separate background from human effects, and it may. But until it does, we have to keep the background effect in mind when we use environmental indices. Figure 3 shows how this effect can be an appreciable part of the index at particular times.

This effect does not exist in all calculable indices. It is zero, as far as we know, for asbestos in the air, cyanides in water, and the cylindrical aluminum flowers by the roadside mentioned above.

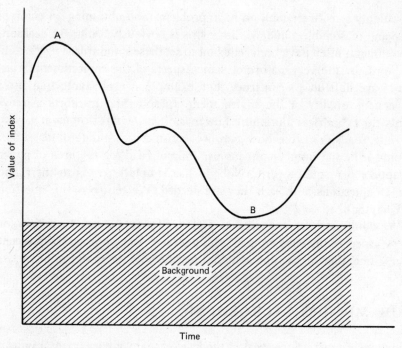

Time

FIGURE 3
*Background effects in environmental indices. Many indices are made
up, in part, of effects that would occur without human intervention.
For example, the amount of dust in the air can increase due to
volcanoes. The number of birds can decrease due to a hard winter.
Humans are responsible for many environmental problems, but not
all. The part played by nature can be called* background. *The diagram
shows the background to be constant with time, though this isn't always so. At time A the human effect on the index is much greater than
the background. It's then mostly our fault, not Mother Nature's. At
time B most of the environmental problems can be laid at her
doorstep.*

ENVIRONMENTAL INDICES AREN'T JUST POLLUTION INDICES

If all types of pollution were stopped tomorrow, we could still have
severe environmental problems. Species of wildlife could be wiped out,
parkland could still be inadequate in many areas, and erosion could
ruin farmland. We can't equate the environment solely with problems of
pollution.

We have to take this broader approach into account when we talk about environmental indices. For example, in a comprehensive index we should consider such quantities as the number of endangered species and recreational opportunities. Because of the vast amount of research that has gone into devising instruments for measuring pollution, data and standards are generally more available for pollution indices even though the environment is more than pollution or its measurement. Indices should reflect this wider view whenever possible.

WHAT ABOUT ENERGY INDICES?

With recurring energy crises and more predicted until the end of the century, news of the amount of oil, coal, and nuclear fuel on hand seems to have pushed environment off the front pages, if not completely out of our minds. But energy and the environment are intimately related. To put it simply, using energy produces a large fraction of the overall amount of pollution in our society. If we use less energy, with all other factors remaining equal, we will have less pollution. If we use the same amount of energy, but juggle oil against natural gas, nuclear power against coal, we will obtain different types and quantities of pollution that affect environmental indices in different ways. For example, a greater emphasis on coal as fuel is likely to produce an increase in an index of sulfur dioxide in the air, and building more nuclear power plants may raise an overall radioactivity index.

Since energy and the environment seem to be so close to each other, why not cement the relationship and marry energy indices and environmental indices? We've chosen not to do so because an index of energy, at least for fossil fuels like coal and oil, is inherently different from our subject. When we say that half of our energy comes from oil, we're not making a value judgment. If, on the other hand, we say that a certain water index is 0.7, we are. The only way that value judgments can be introduced into energy indices is by comparing the amount of fossil fuels we are using to the amount still left in the ground. For example, we might say that at the present rate of use we have 30 years of oil left, 200 years of coal, and so on. These values are indices of sorts, since we are again comparing a measured quantity with a base (amounts left underground). These indices are so different from the environmental indices we've been discussing that it would be confusing to lump the two together. For this reason we won't consider energy indices here, although a good case can be made for devising and publicizing them.

COMBINING INDICES

One possible objection to the use of environmental indices is that we're likely to end up with scores of subindices, leaving the public more confused than ever. For example, eventually we may have half a dozen or so indices on pollutants in air and at least that number for water. Since we don't want lists of environmental indices to be as long as the columns of stock prices in our daily newspaper, we must find some way of combining the subindices.

Fortunately, the problem is one that has occurred before and has been at least partially solved by economists. Economic indices are discussed in more detail in the next chapter, but a few solutions are worth mentioning now. Figure 4 shows one graphically.

For a more numerical approach, let's go back to the fabled land of Gourmetia. In the stories they tell about it, there's more than one type of truffle to be dug up in that nation. Male and female pigs root up different types, as do brown and gray pigs. The varieties differ from season to season. All in all there are dozens of kinds of truffles in Gourmetia, and each has its unique price changes. Somehow an average price index of truffles must be found.

One of the simplest ways of averaging involves weighting. This doesn't involve actually putting the truffles on a pair of scales, but rather giving more numerical significance to the more important varieties. For example, if three times as many black as white truffles are eaten in Gourmetia, black would get three times as much weight in the final calculation. Suppose that each Gourmetian eats an average of 3 ounces of black and 1 ounce of white truffles per day, and that their respective prices are 4 and 2 crowns per ounce. The amount spent on black truffles is then $3 \times 4 = 12$ crowns, and on white, $1 \times 2 = 2$ crowns. The total spent per day is then $12 + 2 = 14$ crowns. The Gourmetians eat a total of $3 + 1 = 4$ ounces per day, so the average price is $^{14}/_4 = 3.5$ crowns per ounce.

One advantage we have had in this calculation of an average index is the use of money as a device for weighting. There is no comparable

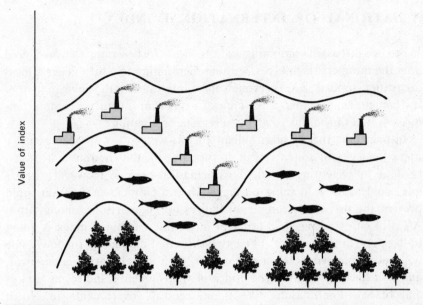

FIGURE 4

Combining environmental indices. Economic indices have a single uniform standard—the dollar—which lets us combine many subindices. Because environmental indices don't have this standard, the mathematics of combination can get tricky. This diagram illustrates how components of an environmental index including air (smokestacks), water (fish), and land (trees) could be combined. The values of each of the parts can vary, but their total is given by the top curve. This method implies that each subindex is added linearly, like a grocery bill. There are other ways, however, of combining these subindices.

standard in the environment. Because of this, we have to use the advice of experts in the field to determine relative weights. For example, it may be concluded that sulfur dioxide should be given a weight twice (or half) that of oxides of nitrogen in computing an overall air quality index. We should remember, in doing this, that (a) it is often difficult to get experts to agree and (b) the further away we get mathematically from the original data and standards, the thinner the ice on which we skate. Only if some of these problems can be overcome can we have environmental indices covering areas such as air, water, and land. Details on what might go into them are mentioned in the following chapters.

A NATIONAL OR INTERNATIONAL INDEX

When we discussed combining indices in the last section, we considered only the numbers themselves, without mentioning to which geographical areas they applied. But everyone's most interested in the environment of his hometown, whether it be Topeka or Toronto. Can we combine indices so that they apply to a city, a region, or a nation?

In principle, there's no problem. Just as we can get an average price index for a city or state, we can do the same for the environment. It can be done by weighting or other mathematical devices. However, there's one problem, now looming no bigger than a politician's fist, that could prevent the use of environmental indices for comparisons among cities. A list of cities arranged in order of increasing combined index is going to have one at the bottom. The city fathers and mothers of the lowly one will not be pleased to be so singled out, even if their rating is deserved, and will likely decry the whole idea of environmental indices as useless and biased. For reasons like these, central governments in North America have been quite wary of listing indices such as that on consumer prices, on a city-by-city basis. The most you can generally hope for, if you want to know whether prices in your town are as outrageous as you think they are, is to obtain the relative change, not the base value of the index. For example, you might be told that the price index for Chattanooga increased by 5% in a given period, but not that the index there was 10% higher than for Nashville.

The only agency with enough gumption to call 'em as it sees 'em on price indices is the usually cautious United Nations, which prepares a set of international indices for living costs of their employees. Since the coverage extends to only one city in each country, though, the indices can hardly be called comprehensive.

Even if we didn't have city-by-city indices, we could certainly calculate national values. This would give us a much better idea of how successful private and public action is in improving the environment. If one of the national indices rose, it could indicate the need for more work on improving that aspect. Conversely, if one went down, we might be able to relax a little.

Eventually it may be possible to design international environmental indices. This would pose even more problems than would national indices, since methods of measurement and standards often vary from country to country. Although the problem looks almost insoluble now, we should remember that definitions and measurements leading to the analogous gross national product (GNP) concept have been converging, in Western countries at least, for many years. For example, the way France defines its GNP is now close to the way the United States does. With a little cooperation the same process can occur with environmental indices. We would then be able to consider the state of the global environment. A lot of lip service is paid to this concept, but until we have some international indices, no matter how crude, we won't be able to make an objective assessment of what the human race is doing to our planet.

NOT JUST NUMBERS, BUT ACTION

Every day, both at work and at home, we're bombarded with numbers and yet more numbers. Adding one more projectile will be about as helpful as putting one more picket on a picket fence unless the indices we compute are the basis for some action. This is not the place to prescribe what the actions should be; they will depend on the indices involved.

For far too long, all of us have been more confused than necessary about the state of the environment. As indices that quantitatively describe it are gradually brought to the public's attention, we should be able to evaluate its condition and then to correct the problems.

2

ECONOMIC INDICES

In September 1974, the Canadian Food Prices Review Board issued a supermarket price index for the Ottawa area. With all the indices being issued by government agencies throughout North America, at first it appeared that this would be the one to break the economists' back. Newspapers seemed bored by it all; one politician roared that because it was an index and was not expressed in good old dollars and cents, nobody would be able to understand it. (His high school mathematics teacher was unavailable for comment.) Bureaucrats at the main Canadian index factory, Statistics Canada, wondered whether the new kid on the block would steal their thunder.

Within a few weeks, the new price index was doing everything that an environmental index should. First, people understood it. Although the price levels were expressed in abstract numbers, shoppers soon became familiar with them and flocked to the stores with the lowest values of the indexes. Second, the Board's numbers produced action. There was actually some competition as stores vied with each other in cutting prices. For example, when the index was first issued, the values ran from 58.33 to 67.80 for the 26 stores surveyed, where higher values indicate higher prices. By the fourth week, the range was from 55.98 to 60.90—a significant reduction in both the higher and the lower range of prices. All this occurred in a seemingly unending era of weekly bouts of depression caused by prices at the supermarket checkout counter.

Because of the rampant inflation affecting almost everything we buy, we can't expect such delightful interludes of lowered prices to go on forever, no matter how well this index works in putting pressure on the stores. But the index *did* work. An index can be more than a number painfully computed by stooped statisticians with celluloid eyeshades. It can and should provoke action to solve some of the problems that it summarizes so succinctly.

14

This book is supposed to be concerned with environmental, not economic, indices. That's true enough, but it's worthwhile considering the latter for at least three reasons. First, economic indices came into being well before environmental indices—by about 250 years. Second, due to this long history of dealing with indices based on money, all sorts of useful techniques for avoiding mathematical pratfalls have been devised. Somebody once said that those who don't know history are condemned to repeat it. There's no need for environmental indices to suffer the disasters that have befallen their economic counterparts. Third, because of the present-day importance of economic indices, there's a veritable army of fact finders gathering and analyzing monetary data all around the world. We can learn how to calculate better indices of the environment by noting how their battalions are deployed.

THE WAY ECONOMIC INDICES CAME ABOUT

Gyrations in the value of money aren't new. One economist has gone so far as to claim that money has been inflated by about 45 times since the Middle Ages. That is, $1 in the fifteenth century would be worth about $45 now. Down through the ages, wars, plagues, and droughts have drastically raised and lowered prices. Philosophers, historians, and other curious people have attempted to find out just how to compare the prices of one era with those of another. The object of their work was to record the changing standards of living just as we would use environmental indices to tell us how we are affecting nature.

Renaissance thinkers realized that the value of money often changed. This was partly because of ordinary changes in the economy, but partly because of the debasement of the value of the currency by monarchs and other nobility through their attempts to make themselves richer and more powerful. But how could these changes be accurately measured?

Probably the first systematic attempt to grapple with these questions came from Bishop Fleetwood in 1707, in his book *Chronicum Preciosum*. Listen to his words:

'tis evident that if £5 in Henry VI days [he reigned from 1422 to 1461] would purchase five quarters of wheat, four hogsheads of beer, and six yards of cloth, he who then had £5 in his pocket was full as rich as he who now has £20, if with that £20 he can purchase no more wheat, or beer, or cloth than the other. . . . You may safely conclude that £5 in the reign of Henry VI was of somewhat better value than £10 nowadays is. In the next place, to know somewhat more distinctly whereabouts an equivalent to your ancient £5 will come, you are to observe how much corn, meat, drink or cloth might have been purchased 250 years ago with £5 and see how much of the modern money will be requisite to purchase the same quantity of corn, meat, drink or cloth nowadays. To this end, you must neither take a very dear year, to your prejudice, nor a very cheap one, in your own favour, nor indeed any single year to be your rule, but you must take the price of every particular commodity for as many years as you can . . . and put them altogether, and then find out the common price, and afterwards take the same course with the price of things, for these last twenty years, and see what proportion they will bear to one another; for that proportion is to be your rule and guide.

This rather lengthy quotation, including that last 97-word sentence, mentions most of the elements used to build modern economic and environmental indices. First, Bishop Fleetwood saw that prices were changing. If they didn't, there'd be no need to measure them continually. The average person doesn't measure the length of his nose unless he's related to Pinocchio.

The second question that the good Bishop tried to answer was how much of a change there had been. He says to combine the prices of corn, meat, drink, and cloth for many past years, and then to compare this number to the equivalent combination of prices for "these last twenty years." In effect, the first number is the standard, or base, which we discussed in Chapter 1. The second number is the quantity in question. The second number divided by the first is the index.

Of course, a quarter of a millenium ago Bishop Fleetwood didn't have all the answers to problems of indices, anymore than a quarter of a millenium from now it will appear that *we* had all the answers. One of the key points he glossed over is just *how* the prices of corn, meat, and drink are to be combined into one number. This is done by *weighting,* which is discussed in more detail later in this chapter. The problem arises because, for example, the price of meat is more important to the cost of living than the price of mustard, and we need some way of taking this into account.

We face analogous problems when we compute environmental indices. Is air more important than water? Is the sulfur dioxide index a more important index than that of oxides of nitrogen? What value can we place on an index of litter compared to other indices? We don't have simple responses to these questions, and cynics might claim that we're still in the era of Bishop Fleetwood as far as this subject is concerned. To some extent this is true, but let's hope that it doesn't take us two and a half centuries to get answers.

Bishop Fleetwood used the quantities of four hogsheads of beer and six yards of cloth, but what relationship did these have to the amounts consumed by typical families in his day? Joseph Lowe, in 1822, was the first to consider this problem. He showed that what were politely called "lower orders" spent different proportions of their income on food, clothing, rent, and the like than their "betters." This fact is taken into account in modern price indices, which are attempts at averaging the consumption habits of everyone. You may love turnips, but your neighbor down the street detests them. In calculating the relative consumption of turnips as compared to cauliflowers in a vegetable price index, the habits of both you and your neighbor, as well as those of everyone else, have to be averaged.

The analogous problem in environmental indices is that different people react in different ways to the same level of pollutants or environmental degradation. If you and I go for a stroll in Los Angeles, the smog level might merely bring tears to my eyes, but it may send you to the hospital with bronchitis. Scientists try to take this into account when they set pollution standards, but so far these guidelines are still crude. It will take more research and thought before the effects of a California walk can be handled mathematically.

The people of Bishop Fleetwood's time may have had to live mostly on corn, meat, and beer, but not too many people do today. Porter, in 1838, was the first to use a large number of subindices (50 in all) to calculate an overall price index. Now price indices are composed of hundreds of the items we buy and services we use.

The problem of "how much is enough" also arises when we compute environmental indices. For example, in constructing an air quality index we might have six subindices dealing with pollutants like hydrocarbons, lead, and the like. How do we know that there isn't another pollutant in the air that we haven't measured, one that would change the value of the total air index?

The truth is that we don't. The situation is a little like finding a lost toy in a dark room—we don't know it's there until we trip over it.

Never being able to say that they're complete might be considered an insoluble problem for environmental indices, until we look at economic indices. They're never complete either. Imagine yourself at your local supermarket, with thousands of different items on the shelves. There just aren't enough people to go around checking the prices of every article in every store. If there were, there wouldn't be enough people to produce the articles in the first place. So economic indices have to be selective too.

The problem of incompleteness is solved by a technique called *correlation*. It turns out that the prices of some articles are strongly tied to (or correlated with) the prices of other articles. For example, the cost of frozen orange juice is always highly dependent on the cost of fresh oranges, so there's no need to measure the price of both. In this way we can calculate economic indices on the assumption that a relatively small group of quantities can stand for the tens of thousands of articles and services that a nation pays for.

Research is now under way to determine which environmental quantities correlate with others, so that we don't have to measure everything in sight or, as in the case of air pollution, everything not in sight. It's too early to tell how this work will come out, but we can assume that lack of completeness won't be a drawback to environmental indices.

But back to history. After the pioneering attempts at constructing the economic indices mentioned above, it seemed as if everyone would get into the act. Long mathematical books were written on how to combine indices in the best way. Unfortunately, the authors never agreed with each other, and what we use now is a compromise among a number of viewpoints. Like the perfect chord in music, the perfect formula for combining indices on a year-to-year basis has yet to be found. This problem doesn't seem to have stopped anyone from advocating his own set of indices.

During the middle and late nineteenth century, the demand for indices became so great that many business magazines and economists produced their own series. An analysis of the good and bad points of each would fill volumes, and undoubtedly did fill with confusion the heads of those who were attempting to manage national economic policies. Governments had to set up some sort of official price index. The history of national indices varies from country to country, but in the

United States efforts go back to World War I, when the Bureau of Labor Statistics participated in studies of living costs in shipbuilding and railroad centers. By 1919, a general cost of living index had been issued by this bureau. This index has changed its name a few times and had its basis of comparison revised frequently since 1919. Work in other countries usually started later, but by now almost every national government publishes some type of consumer price index.

We've been talking primarily about price indices, mainly because they're so common and their history is so well known. But there are many others, both official and unofficial. Many of the latter are issued by magazines concerned with economics, and are narrower in scope than those of the government, whose tax dollars allow a more thorough treatment. We'll only mention a few of them in passing, since we're more concerned with protection of the environment than of the economists. Some of the official and unofficial indices include the Dow-Jones index of stock prices, indices of hours worked and wages earned in manufacturing, the U.S. Department of Commerce foreign trade indices, indices of currency circulation, many types of industrial output indices, and bond price indices.

Such concepts as the gross national product are examples of indices, and these are discussed in some detail later in this chapter.

We can see that economic indices have made the way much easier for environmental indices. However, even economic indices are still far from perfect. As a result, we shouldn't fret that environmental indices aren't problemfree either.

THE DOLLAR—THE ALMIGHTY STANDARD

When we talked about standards for environmental indices in Chapter 1, we noted that devising them would sometimes be difficult. We have to somehow juggle health, ecological, and aesthetic effects to come up with a number or set of numbers to serve as standards.

The makers of economic indices seem to have an easier time of it. They have the dollar (or pound, or mark), and everything they work with can be computed in terms of it. When you have a rigid and uniform yardstick, you don't have to worry about devising other standards. Trouble arises when the yardstick has different lengths depending on when we pick it up. The reason for this is that the dollar has varying values as time passes. We generally know this phenomenon as inflation—the decrease in the value of money—but occasionally there have been periods of deflation, or an increase in its value. This would be rather pleasant except that these periods are usually accompanied by depressions.

So in one respect, environmental standards are more stable than economic standards. The concentration of a given air pollutant generally bears a definite relationship to human health. But a standard whose value has decreased by a factor of 45 times in the last five or so centuries is obviously a yardstick of a different color.

Hold on, cry the economists, things aren't as bad as they seem. It doesn't really matter that the yardstick keeps shrinking, they might claim, as long as we know its relative length all the time. That's what indices like the consumer price index attempt to measure.

There *is* some truth in this argument, but only some. The problem is that we really don't have only one shrinking monetary yardstick. We have many, and they're all disappearing at different rates. For example, we're accustomed to thinking of the consumer price index as the sole measure of inflation, or the rate of dollar shrinkage. But another inflation measure is used to adjust the GNP for changes in the value of the dollar. (The GNP is discussed later.) Similarly, other economic indices are preshrunk in different ways depending on how they're calculated. It's true that the "shrinkage factors" used to fix up the various economic indices aren't drastically different from each other, but they aren't the same.

Using the dollar as a standard for economic indices is vastly superior to using any standard we can devise for environmental indices, because the dollar is as good as gold. The only question is, what's the price of gold? And what was it yesterday?

TWO TYPES OF ECONOMIC INDEX

Economic indices aren't confined solely to measuring price changes of consumer goods, commodities, or stocks. There's a much more sophisticated type of index that is an attempt to measure the overall state of the economy—the gross national product. The GNP is supposed to be the measure of the value of all goods and services produced in a country. Strictly speaking, it isn't an index, since it's expressed in terms of those rubbery dollars we talked about in the last section. However, most people who haven't seen anything larger than a $100 bill have trouble visualizing the billions and even trillions of dollars in which the GNP is expressed. As a result, the GNP is often talked about as if it were an index—for example, this year's GNP might be described as 10% higher than last year's, or 25% higher than that of three years ago. In such an index the GNP of last year (or 3 years ago) is the standard, and the measured quantity is the present value of the GNP.

The GNP uses economic data, but combines more of it than the other indices we've talked about in this chapter. Although the GNP has by now assumed the role of a sacred and mystical number, it's calculated by adding and subtracting everyday quantities. Since this book is not intended to be yet another weighty and unread tome on economic theory, there's no need to go into great detail on why some are added and others subtracted. Added in are quantities like wages, salaries, and bonuses; indirect taxes such as sales taxes; corporation profits before taxes; interest and other investment income; and net income of farmers. Some of those subtracted are dividends paid to foreigners and the dollar value of the depreciation of buildings, factories, and machinery.

In spite of all the calculations that go into it, the GNP has digits of clay. First, all unpaid services are deliberately excluded. For example, volunteer labor and the work of housewives are left out, not because they are unimportant, but because they don't involve money. Many more transactions have value, but not cash value, and the GNP doesn't consider them. Second, and more important from the point of view of those concerned about environmental degradation, the GNP doesn't differentiate between "good" and "bad" activities. Suppose we have two factories, side by side. Each disburses the same total payroll, coughs up the same corporation taxes, and pays the same sales taxes on the materials it uses. In addition, the depreciation of the building and the machinery it contains are the same for the two factories. But suppose one puts out tons of water and air pollutants each day, and the other, through ingenious methods, keeps its pollution down to the absolute possible minimum. As far as the compiler of the GNP is concerned, the two factories are exactly the same, but in the view of those of us concerned with the environment, they couldn't be more different. Figure 5 shows how the calculation comes out wrong if we don't take the environment into account.

All of this is not to say that the GNP has no value. It's the most

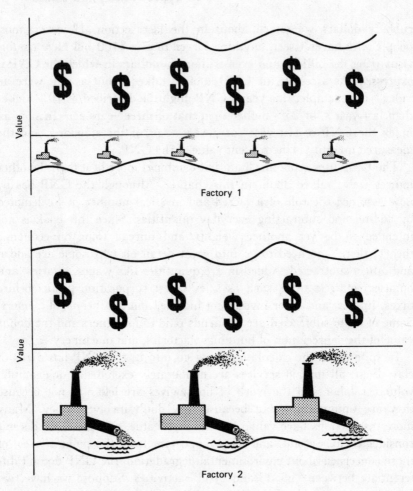

FIGURE 5
Comparison of the amounts of pollution generated by two factories. The dollar signs represent the value of the goods turned out by the two factories—both equal. Below the line is the monetary value of what economists would call disbenefits, *or the harm the pollution causes. The gross national product counts everything above the horizontal lines and ignores everything below them. To the GNP the two factories are equal. Or are they? Judge for yourself.*

comprehensive and sophisticated economic index we have. But in spite of all the work that goes into its compilation, it still doesn't tell us everything we want to know. We need an analogous set of information to tell us something about the nonmonetary aspects of life, and environmental indices will eventually fill part of this gap.

The GNP could be described as a cumulative economic index, as opposed to a simple economic index like consumer or stock prices. That's one of the reasons that it is so valuable. Can we have an analogous cumulative index for the environment?

We can, but it's going to take much more knowledge than we have now. Chapter 1 briefly discussed how to combine separate environmental indices. The discussion was rather short because there are so many areas on which people disagree. The prime problem is what we mean by the total environment. This is the concept that would be the ecological equivalent of the GNP. When you say *total environment,* you may really be stressing wildlife and parkland, while your neighbor may be almost totally concerned with the levels of air and thermal pollution.

There's still some disagreement among economists on how the GNP should be calculated. Some may want to exclude certain types of income and include others not now combined. The levels of difference among highly trained economists is naturally going to be less than that among those concerned with the environment, which includes almost everyone to at least some degree. Because of this disagreement, later chapters suggest some ways in which environmental indices can be combined to form some sort of analogy to the GNP, but it will only be that, a suggestion. In the years to come, there are inevitably going to be better ideas on how to do it.

WEIGHTING OF INDICES

Suppose that you were an economist going out to gather numbers to put into a consumer price index. At your local meat market hamburger is $1.00 a pound and round steak is $1.50 a pound. (These prices are a bit of fiction in a nonfiction book. In these days of rapidly rising prices, any number used is bound to be out of date rapidly.) How can these two prices be incorporated into an overall consumer price index?

It's done by *weighting.* In other words, each quantity that goes into the price index is given a weight, or a number denoting its relative importance compared to everything else in the index. The simplest way to explain this concept is with an example from a chart of weights of an old U.S. consumer price index. In this chart, food was assumed to account for 29% of all spending, or have a weight of 0.29. In the food section, round steak had a weight of 0.0079, hamburger 0.0043, bacon 0.0092, apples 0.0005, and so on. To find a total price index, we multiply each of the weights by the price of the article, and add up all these terms. Figure 6 shows weights for one possibly typical person.

In Chapter 1, we used the example of Gourmetia when we introduced the idea of weights. The lucky inhabitants of that land live only on champagne and truffles, so computing their price index was easy. However, the chart used for the numerical example mentioned above has hundreds of weights, as a sample of the thousands of things we buy and use.

It's a little unfortunate that the word *weight* means both something that makes a scale tilt as well as a mathematical quantity. A hammer weighs more than a tiny bottle of spices, but the mathematical weight of the spices in the price index may be more than that of the hammer. The English language is sometimes all Greek.

When you read down the list of weights assigned to each item in a price index, you may think that because they're carried out to four or five decimal places, they're extremely precise. They're not as accurate as they seem, and the next few paragraphs explain why. This is done not to destroy the public's faith in economic indices, but rather to point out that any index is not perfectly accurate. One of the objections sometimes made against environmental indices is that many of the areas with which they deal aren't understood fully. True enough, but it also applies to economic indices.

Let's get back to the supermarket we were visiting a little while ago. It's too bad the economist came by today, because yesterday they had a great special—hamburger at $0.80 a pound and round steak at $1.25. The meat counters were so jammed with shoppers that you could hardly get at the bargains. Today, with meat at the regular price, the place seems deserted except for the economist poking around. How can he (or she) take account of yesterday's prices when he wasn't even in the store? He can't, so only today's higher prices get recorded and sent to headquarters. We can't have economists or statisticians stationed at every store every hour of the day. Therefore, there are bound to be some errors in any price index.

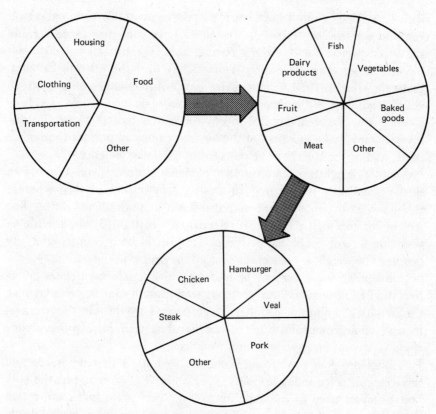

FIGURE 6

*A rough example of my spending habits. Index weights are a measure
of the relative importance of different parts of an index. In a price
index they indicate how much income is spent on various outlays.
Each of the components in the upper left-hand circle can be divided
into subcomponents. For example, food outlays are used to buy meat,
vegetables, milk, bread, and dozens of other products. The sizes of the
wedges demonstrate the average amount I spend on each, or their
weights. The process can be carried further by breaking down each of
the components in the upper right-hand circle. We can use this
hierarchy of weights to build a complete price index.*

What does this have to do with weights? The weights are supposed to
be an indication of what the average person or family buys. Most people
are going to buy bargains if they can. If these bargains are missed by the
economists, they won't have as good an idea of what people are buying.

A more serious problem lies in the fact that what people buy changes with time. You couldn't have bought a television set 30 years ago even if you had wanted one. Expenditures on ice have gone down considerably since that period due to electric refrigerators. Weights allocated to television and ice must have altered drastically over the last few decades. Since the weights used in most price indices are changed only every 10 years or so, they're almost always a little out of date.

The same problem applies to environmental indices. Scientific discoveries may make one part of the air index more important than other parts, and this will have to be reflected in changing weights.

Another weighty problem is that of regional differences. Way down in the boondocks, people may sit around drinking tea all day, whereas sophisticated urban dwellers may detest tea and stick with coffee. The weight for tea in the chart we referred to was 0.0013, but there's no way one weight could reflect the preferences of both country and city people. The weight is an average over all the people in the nation.

In a similar way, you may be more sensitive to a form of water pollution than I am, and in your personal water quality index you may give the irritating pollutant a higher weight than I would. On the average, though, some compromise is reached between your, my, and everyone else's viewpoint on environmental weights.

A final problem concerning economic weights is that the goods and services in a price index are only representative. For example, the only beef products used to calculate an average beef price index are round steak, rib roast, chuck roast, and hamburger. There are at least a dozen other cuts available from even the small meat counters, and more from specialized grocers. The four mentioned are supposed to be indicative of all the others as far as price changes are concerned.

Air indices may not yet measure everything that mankind puts into the air. If they are designed appropriately, they can be representative of indices that are not yet computed.

It might be more satisfying to our ideas of completeness to measure all beef costs for a price index and all air pollutants for an environmental index, but there's a limit to how much time and money we can spend on either. We'll have to be satisfied with partial results in both cases.

There are obvious approximations made when we calculate mathematical weights for price and other economic indices. This hasn't stopped us from going right ahead and using them, and shouldn't when we turn to environmental indices.

QUANTITY AND QUALITY

Your hometown may have two or more newspapers. If it has only one
or even none, you probably have access to those from out-of-town, from
the capital, or even the few national newspapers. If you're like most
people, you probably buy only one paper each day. Why don't you take
two, three, or more? You may find the others boring, with too much or
too little sports or fashion news. In fact, there may be dozens of reasons,
both big and small, why you choose one newspaper over another. To
put it another way, you believe that your paper's *quality* is higher. Let's
suppose that all the newspapers you can buy are the same price. To the
economist they're all the same because each has the same price. You
may protest that your paper has your favorite cartoon strip, and none of
the others do. But in calculating his economic index, money is what
counts, not your odd preferences.

Quantity is something that economic indices have no difficulty with;
quality is another story. To partly solve this problem, the makers of
these indices have set up "specifications" for the items they use in the
hopes of eliminating quality differences as much as possible. For
example, going back to the old price index we've referred to, the specifications for men's street shoes comprise eight details. The upper is to be
"kip side, good quality (grade)"; outsole, "leather, No. 1 scratch, 8 to 9
irons"; insole, "grain leather, gemmed, 4 to 5 irons." Precisely what
these specifications mean may be a little difficult for a nonshoemaker to
determine. Even after these specifications are met, there could still be
differences in quality. For example, the specifications for motor oil read
simply "regular." It's well known that there are some brands of oil that
motorists swear by, and others that they swear at. As far as the price
index is concerned, they're all "regular."

The question of quality is one in which environmental indices can be
better than economic indices, because many are based on physical
measurements. Generally speaking, a physical measurement is
inherently more accurate and reliable than an economic measurement
and doesn't depend on man-made definitions of "quality." Physical
measurements are not subject to many of the problems we've discussed
in this chapter.

It's true, though, that some environmental indices can depend on
quality judgments. For example, aesthetic indices, which are discussed
in Chapter 8, are based on what each person thinks is a beautiful land-

scape. As one of the section headings in that chapter has it, "Your yard is messy, mine is only lived in. "If we want to overcome personal bias in this group of indices, it will take more research. To sum up, the problem of quality is one that bedevils most economic indices. In this area environmental indices can be more accurate.

NOBODY IS AVERAGE TO AN INDEX

A rather strange family, Mr. and Mrs. Average with their 2.3 children, stalks the halls of the office buildings in which economic indices are calculated. At first glance this family might appear quite unexceptional, except for that three tenths of a child. They smoke a moderate amount, drink typical quantities, and eat foods that are about average. The house they live in is not too different from other people's, and they trade in their car every 3 years or so, just like most of the rest of us.

There's just one problem with the Average family—they don't exist. Look around at your friends and relatives. Some smoke much more than the average, and others refrain completely. The ones that are average smokers may drink more or less than the average. The ones that appear normal in most respects may have odd habits you don't even know about, like existing solely on champagne and truffles.

Nobody is really average in all respects. Therefore, the consumer price index, which is based on average buying habits, applies to nobody in particular, and yet to everyone. For example, I may become convinced after studying some mystical religion that the best way to live is by consuming only raw cabbage. If the price of cabbage doubles, this will alter my cost of living tremendously, but the overall price index will be insignificantly affected. In the same way, price changes touch everyone differently.

The same problem occurs in environmental indices. A particular level of water pollution may bother me so much that I leave the beach, while you go on happily splashing away. All we can say is that every index is an average over people's feelings and responses.

IS AN ENVIRONMENTAL INDEX MORE DIFFICULT TO PRODUCE?

An environmental index shouldn't be more difficult to produce than an economic index. We now have masses of environmental data that can be used. Although we don't have quite as good geographical and time coverage for the environment as we do for the economy, this is rapidly changing as governments set up more and more monitoring stations. Environmental indices have one great advantage and suffer one great disadvantage as compared to economic indices. First, the good news. Many of the environmental measurements from which we obtain indices are based on well-known physical, chemical, and biological principles. This avoids many of the subjective and personal biases, especially in the realm of quality, that economic indices can suffer from.

Now for the bad news. Over the past half century, economists have built up a strong theoretical base for the indices they use. They now have a fairly good idea of the limitations of the index concept, and where it can be improved. In addition, they've convinced governments of the need to compile their type of index, with the result that thousands of economists around the world are employed analyzing and refining economic indices.

Environmentalists are not as fortunate. Most professional environmental scientists have confined themselves to fairly narrow fields, and comparatively few have tried to integrate all the accumulated knowledge and data. In part, this is an attempt to do so. Environmentalists won't ever reach the relative sophistication of economists until they, and we, put considerably more effort into concepts like environmental indices.

An environmental index isn't really more difficult to produce than an economic index; it takes different talents, but the same hard and continuing work.

THEY AREN'T SO PERFECT AFTER ALL

Not everyone is convinced of the necessity or usefulness of economic indices. One person has written:

> With the exception, perhaps, of stilts for a serpent, there is nothing more useless or ridiculous than a mass of figures collected at great travail, added, multiplied, divided by the cube root of π and converted to homogenised index numbers, that have no bearing on the problem.

Most of us think that these indices can be helpful in understanding the state of the economy, provided that we know their limitations. Unfortunately, we often tend to forget these limitations. The object of this chapter was not to show that economic indices are valueless, but that they're not perfect. To produce them, a horde of assumptions must be made. Since many similar assumptions have to be made to produce environmental indices, the two concepts aren't so different after all. We can use the experience that economists have gained in building their indices to produce better indicators of the state of the environment.

3

WHAT'S BEEN DONE SO FAR?

Concern for the environment has been around for a long time, under such terms as nature study and conservation of resources. Indices measuring its state go back many years as well, although they've lurked more in the background of public consciousness. This chapter examines what has been done in the past, and looks at both the good and bad points of this work.

THE FIRST ENVIRONMENTAL INDEX—HEALTH

Someone strides out into the fresh morning air, and is soon coughing and sneezing. The air isn't as fresh as he had thought. Someone else takes a refreshing drink of water, and a few hours later feels his stomach trying out for the high jump championship. At first, we might think that all we're seeing are a few of the symptoms that a doctor encounters every day, but there's more to it. The coughing, sneezing, and stomach flip-flops are really crude forms of indices—in fact, the first such measures of the state of the environment. Figure 7 shows graphically how the state of health can vary with the state of air and water.

Long before we had complicated instruments or involved mathematics to convert their readings into indices, people suffered from pollution. The degree of human suffering can be called a simple and early form of an environmental index. Of course, nobody could measure pain, but as people crowded together in cities and the Industrial Revolution began, more and more had health problems attributable to smoke and the strange substances in the water.

Because there were so many people dying of all sorts of diseases in those days, there was no means of knowing which deaths were a result

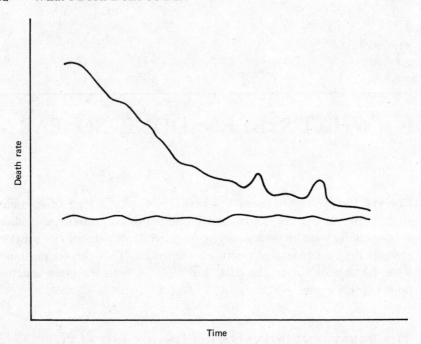

FIGURE 7
Changes in death rate for a disease associated with environmental conditions, such as bronchitis. The upper curve represents the overall death rate. The lower curve shows the rate that would hold if environmental conditions were uniformly good; in effect, a type of background. The difference between the two curves shows the deaths that can be attributed to a polluted environment. The decreasing difference with time implies that, for this disease at least, the state of the environment is getting better.

of what we would now call environmental causes. Nonetheless, they did occur. Medical science is now able to diagnose the diseases, such as bronchitis, that are caused or aggravated by pollution. Environmental science has devised many instruments by which we can get a more accurate analysis of our air, water, and land.

When environmental conditions are bad enough, health still forms the basis of an index. For example, the outbreaks of heavy air pollution at Donora, Pennsylvania, and London about a quarter century ago killed many people. Measurements were taken at that time of the physical level of pollutants in the air. By using the relationship between these levels and the incidence of respiratory problems, we can devise a rough scale for an index of air pollution effects. In fact, this very information is used in the Ontario Air Quality Index, which is discussed later in this chapter.

Now that we have, and are continuing to develop, some fairly so-phisticated ways to measure environmental quality and levels of pollu-tion, we don't have to rely solely on human health as an indicator of the state of the environment. One of the objects of this book is to discuss how indices can at least partly replace the human body as guides to this knowledge. For centuries our bodies served as large guinea pigs in tests of how much environmental degradation could be tolerated. As we move into a more precise era of numbers, we should keep in mind these first primitive indices.

ALL THE INDICES FIT TO PRINT—THE NEWSPAPERS

The origin of the concept of environmental indices is unknown. However, regardless of all the dry debates in academic journals about who said what to whom, it seems clear that newspapers were the first to bring the simpler ideas behind indices to the public's attention.

Ralph Waldo Emerson once said that the newspaper "does its best to make every square acre of land and sea give an account of itself at your breakfast table." If these acres could speak, one of their complaints would be about the treatment they have received from mankind. News-papers have, over the past decade, given them voice by issuing forms of environmental indices.

Most of these indices were fairly crude and usually dealt with only one aspect of the environment, such as air pollution. They had the ad-vantage of quickly spreading the index to a wide audience. They had the disadvantage that the scales used were sometimes difficult for many to understand. In addition, each newspaper that issued indices seemed to have its own peculiar way of defining the numbers it used. Certain regions had a variety of environmental indices in the local newspapers, with some increasing and others decreasing as pollution levels increased. When the information seems contradictory, many people shut their eyes and ears to all of it.

Newspapers are by nature transitory. Environmental indices can be used as a means of getting ecological news hot off the press, but if that's all they're used for, they become just a string of vaguely connected numbers. The knowledge has to be put in context by the calculation of trends and the effect of these trends on the state of our surroundings. Living day-to-day, newspapers are usually not equipped to or have the inclination to look at the broader, long-term view. Although the daily press was the first to bring environmental indices to the eyes of many, getting them in the paper is only part of the job.

THE NATIONAL WILDLIFE FEDERATION INDEX

In the late 1960s the National Wildlife Federation, the largest conservation organization in the United States, started its E.Q. (Environmental Quality) Index. Each year since then, similar indices have been published in its magazine, *National Wildlife*. This was the first attempt to gain a national picture of the state of the environment through numerical indices.

Accompanying the E.Q. Index was a *Reference Guide,* which listed the sources of data, as well as the relative importance of the subindices that made up the total index. For example, the 1971 Index had seven components: soil (with 0.3 of the total), water (0.2), air (0.2), living space (0.125), minerals (0.075), timber (0.05), and wildlife (0.05). Soil, water, and air have about two thirds of the total value, with the other parts accorded less significance. The index scale is from 0 to 100, the former signifying complete environmental degradation, and the latter perfect conditions.

In the eyes of scientists, the main problem with the E.Q. Index is that the way in which the values of the seven subindices are calculated is not made clear. For example, in the 1971 index the air subindex had a value of 34 (or about two thirds of the way to complete degradation), the timber subindex was 76, and so on. These numbers were not determined by means of equations, but by the collective judgment of the magazine's staff. Each issue of the index makes clear that the values are subjective.

Equations are not necessarily better than human judgment. Because equations are devised by humans, they are always subject to the errors of their creators. Equations do have two advantages over getting a few people together in a room to devise the value of this year's index: They force a person to put all of his assumptions on the table, and they don't change spontaneously, as the human mind often does.

For example, environmental experts can and do disagree. The assumptions that go into their reasoning are very rarely spelled out in black and white. However, if our equation for air pollution states that carbon monoxide is equal in importance to, or half or twice as important as, sulfur dioxide, the mere writing down of the mathematical weights requires us to reveal the basis of our thought processes. It is true that equations can be and are modified as we learn more about the environment. If they weren't, we probably wouldn't be learning as much as we should. But equations are relatively constant in assumptions, and this is what we seek in environmental indices.

Even if those who devise a subjective or nonmathematical index were the most highly respected experts in their field (remember the saying that an expert is an ordinary person visiting another town), it would be difficult to ensure that this type of index is a valid representation of the environment. The staff of any magazine changes, and *National Wildlife* is no exception. In consequence, each year the viewpoints used to formulate the index are not the same as those of the previous year. Year-to-year comparisons, one of the aims of a true environmental index, are relatively invalid.

The E.Q. Index is still one of the few attempts to evaluate the environment of an entire nation. It has the advantage over newspaper indices in that it is an attempt to take into account a wide range of subindices, instead of only a few of them. Because of its defects, scientists were eager to improve their understanding of the advantages and limitations of indices. This naturally led to one of the old reliables of scientific meetings—a symposium.

AAAS SYMPOSIUM

The American Association for the Advancement of Science (AAAS) is the largest scientific organization in the United States, if not the world. When, at their 1971 annual meeting in Philadelphia, one of the subjects that a large number of scientists gathered to talk about was environmental indices, the field had clearly reached a take-off point.

About twenty papers were presented at the symposium, and they were subsequently published as a book, *Indicators of Environmental Quality* (W. A. Thomas, Ed., Plenum Press, New York, 1972.) Although many of the papers were highly technical, others were of a simplified nature. Since the symposium was a landmark in the field, some of the results are discussed elsewhere in this book. Here we can give a broad overview of the scientists' attitudes.

Several speakers evaluated the ways environmental indices fit the pattern of measurements of how we live. Many felt that today's economic measures are not enough. We need what can be termed *social indicators,* of which environmental indices are an important part. How the latter indices relate to others, such as aesthetics, crime, and living space, concerned many of the scientists. Chapter 9 discusses this.

Another general point concerned who would use indices. Chapter 1 mentioned the likely groups. The AAAS meeting showed that planners—on a local and regional scale—would probably be among the greatest utilizers.

As might be expected from a group of scientists from many disciplines, most of the papers concerned only particular aspects of environmental indices. For example, separate discussions concerned the use of water plants, animals, and fish as indicators and indices; the use of land plants as indices of air and land pollution; ways in which biochemical tests on living and dead animals can be used; smell and odor, noise and sound; soil quality and measures of radioactivity.

The result of the symposium was not sweet agreement on each and every aspect of environmental indices. After all, there have been a few international diplomatic conferences in which unanimity wasn't reached. However, the papers show that the majority of the experts present felt that, at least in their specialty, indices could and should be constructed.

Most scientists at the AAAS Symposium knew quite well the relations of their fields to indices, but had only a nodding acquaintance with other specialties. In science, as in cobbling, sticking to your last is the approved way. The problem is to combine the knowledge of many disciplines into an overall approach to environmental indices—a difficult task. The next section discusses attempts to calculate combinations of indices.

THE ROLE OF INDEPENDENT GROUPS—MITRE AND OTHERS

The report of the MITRE Corporation on environmental indices could be called a coffee table book—if the legs of coffee tables were strong enough to support it. The report's 920 pages covered the subject exhaustively and perhaps exhaustingly.

Since we obviously can't discuss all its details, only the major points are mentioned here. The report originated in 1971 as a background study for the Council on Environmental Quality (CEQ) in the United States. MITRE was asked not only to outline ideas for indices, but to say which should be used and why. Many of their suggestions are still valid, and some are already being employed.

Because governments will be the eventual major index-makers, why were private and nonprofit organizations brought in to evaluate the needs for these indicators? The answer is simple—officials are often cautious about issuing new sets of information to the public until the ideas have been evaluated by an outside group. The outside jaundiced eye is sometimes worth its weight in oil.

A total of 112 separate indices, ranging from odors in the home to fish kills, are considered in the MITRE report, and equations are given for the detailed calculation of each. Many of the indices are hypothetical; that is, little or no data exist that could be put into the form of indices. In other cases, real data are available, and the report shows the ways it could be used.

A few of the proposed indices are quite technical, such as some of those discussed in later chapters on air and water indices. Others are more fanciful, such as the proposal to produce a litter index by counting the number of bags of garbage that Boy Scouts fill per mile of waste-strewn highway.

The proposed indices are arranged in 14 broad categories. Because there are only so many ways in which one can subdivide the environment, many of the categories are similar to those that are used in this book; for example, land, radioactivity, and the like. Others deal with subjects that are usually considered only partly environmental, such as resources, the weather, and population. Finally, some categories of indices consider monetary aspects, such as economic losses due to pollution and the amounts spent on pollution control and research. This book has comparatively little to say on the relation of money to the environment, for two reasons. First, the subject has been worked over in scores of books, although not from the index point of view. Second, money spent on improving the environment doesn't always produce results. Since we are attempting to measure actual improvement, we need real scales, not dollar scales.

An important part of the MITRE work is the attempt to determine which of the 112 potential indices should be issued first. The ranking was done by calculating the ratio of the estimated importance of the index to the cost of gathering the data that would go into it. For example, indices of the amounts of pollutants discharged into the air were judged to have high value (in terms of understanding) and would cost relatively little to produce, so this index is one of those recommended. Conversely, an index of odors in the home would be of comparatively little value and would cost a great deal to produce, so this index stands low on the scale of recommendations.

This book does not try to rank potential indices according to this rough "benefit–cost" ratio for two reasons. First, doing so would add yet more value judgment to a field in which the basic value judgments are not all agreed upon. Second, the desirability of different types of indices varies with the particular region, state, province, or country being considered. Eventually, each nation must decide which environmental indices it wants.

The MITRE study provided valuable insights on scaling of indices, their combination, and a host of other problems that are faced by index makers. A careful review of all the work might require a report as large as the study itself. It's enough to say that the study will prove to be a mine of information for a long time.

The MITRE work was primarily concerned with a national approach to indices, but they can also be produced on a smaller scale. In the United Kingdom, the Pollution Research Unit at the University of Manchester recently published a volume, *The Geography of Pollution,* that evaluates the state of the environment in Greater Manchester. The metropolitan area is about 500 square miles and contains a population of about 2¾ million. Many of the techniques and calculations used can be applied to cities throughout the world.

At 150 pages, the Manchester book is somewhat shorter than the MITRE work, and so is a little more approachable. The authors gathered as much environmental data as they could on the 71 sections of the region, and combined most of it into six indices. Three of the indices deal with air, (smoke, sulfur dioxide, and automobile density), two with water (river and canal quality), and one with land (land pollution). The last-named index is based on the fraction of land in each of the sections that was occupied by refuse and garbage dumps.

The six indices were added together in a simple way, so that the relative environmental quality of the 71 sections could be ranked. As might be expected, sections closest to the city center had the highest value of the composite index (or lowest environmental quality), but some suburban sections also had high indices.

Since the Manchester study was concerned with understanding as much of the regional environment as possible, the authors presented much data that could not be put into the form of indices. For example, some indirect information existed on noise pollution, but not enough to make an index. Nonetheless, all available information was carefully discussed, even that not in the form of indices. Indices are not the only tool for assessing the state of the environment, although they often prove to be one of the more important ones. Information may not fit into the mold, and yet should be presented to the public.

The Manchester study can serve as a model of index presentation for small regions. It is clear, informative, and short—attributes that are rarely found together.

Many other groups and individual scientists have tackled the problem of environmental indices. The CEQ in the United States commissioned a number of studies less wide-ranging than the MITRE study. For example, Enviro Control, Inc., studied water indices for about 140 monitoring stations, considering such factors as dissolved oxygen, total dissolved solids, cloudiness of water, and nutrients. The values for each parameter were compared to the previous year's readings, and that aspect of water quality was then judged to be better or worse. This is, of course, a crude form of an index.

Another CEQ study, by Earth Satellite Corporation, deals with land use indicators and indices. The Stanford Research Institute published one on pesticide indicators, as did the Smithsonian Institution on wildlife. Details are discussed in later chapters.

Not every piece of work has been done; each study points out wide areas of environmental ignorance. Yet environmental indices have been evaluated from literally dozens of points of view, and most independent scientists who have worked in the field have urged that a start be made in producing them. Their efforts for the past decade have laid the groundwork for the indices to come.

DOES A GOVERNMENT DARE?

Most governments have been relatively reluctant to put their mouths where their money is. They don't mind subsidies for index research; but actually carrying out the recommendations resulting from the work is another matter.

There have been some exceptions to this rule. Some governments have issued environmental indices, but these have generally been confined to only one small part of the environment. Although politicians are fond of claiming that they "look at the big picture," this hasn't extended to a picture of the environment.

Let's look at the halting steps that have been taken. As might be expected, the first environmental indices dealt with air pollution. The reasons were mentioned earlier in this chapter—the pollution is visible and the measuring technology is fairly standard. As in almost every situation since Adam and Eve, its hard to say who was first. However, some of the first indices were issued by American cities such as New York, Detroit, Buffalo, and San Francisco. Each used a formula differing slightly from the others, but most included the two most obvious pollutants—sulfur dioxide and particulate matter in the air. (The level of particulate matter is measured in different ways, and an analogous measurement is sometimes called coefficient of haze.) Others measured such pollutants as carbon monoxide, nitrogen dioxide, and oxidants.

It was perhaps inevitable that each city producing air quality indices would have a slightly different approach from the others. When it comes to indices, the situation sometimes resembles the days of Greek city-states, when each town had its own system of government. Fortunately for the concept of indices, the differences are more cosmetic than real. City A may have an instrument for measuring carbon monoxide, and city B none, with the result that this pollutant isn't included in B's air index. But the scales that the two employ tend to be similar and, as different types of instruments become more common, indices will resemble each other more.

A nation consisting of city-states would nowadays be a strange political entity. The trend towards amalgamation also exists in the realm of indices. Knowing that the index has risen or fallen in Buffalo or Detroit gives us a clear picture of air quality in two cities, but not over a wide region. Perhaps the first environmental index to combine data from many cities into a standardized scale was the Ontario Air Pollution Index, first calculated in 1969.

At the time of formulation of this index, photochemical smog very rarely occurred in this Canadian province. In consequence, the pollutants that produce it, such as oxides of nitrogen and hydrocarbons, are not included in the index, which evaluates only sulfur dioxide and particulate matter. These are the two pollutants most commonly considered in city-wide air indices. The scales and weights used to produce an Ontario index value are complicated, as is the reasoning that went into their derivation, and therefore won't be repeated here.

The Ontario air index has been used primarily as a legal device and public warning system, rather than as a means to determine environmental trends. For example, an alert is called when the index reaches a certain level, and some polluters may be ordered to stop. A second public alert is sounded when the level reaches a higher value. Finally, when an even higher index indicates that there is a potential "air pollution episode," the government can shut down all sources of pollution "not essential to public health or safety." In these ways, the index is used to signal the quality of the air and their resulting responsibilities to both the public and regulating agencies.

Other regional political units are working on other aspects of environmental indices. For example, Arizona has under consideration the Arizona Trade-Off Model (ATOM) Plan. Because of its complexity, we can't fully describe it here, but essentially a system of indices is used to gauge whether industrial development is appropriate in certain areas.

These last two examples have indicated that environmental indices have far more uses than just telling us how we're treating our surroundings. They can be used by law enforcement agencies to order a reduction in pollution, and by planners to decide where industry should be located so as to harm the environment least.

The logical step after consideration of the state and provincial levels is to see how countries have handled the problem. National authorities have tended to be more cautious than have those in lower levels of government. There are reasons for this. A nation usually covers territory having a wide range of environmental conditions, but most indices consider these conditions from a standard point of view. As a result, some national governments consider that such indices don't take enough account of regional differences and are therefore reluctant to evaluate the environment on a country-wide scale.

Let's illustrate these vague statements with a few examples, and then see how the problems might be overcome. To produce a coefficient of haze (COH) index, polluted air is blown through a filter paper and light is shone through the soiled paper. The more smoke and dust in the air, the less the light shines through. The light can be measured and compared to a standard. Regional differences enter in because the color of the air particles varies from city to city and from area to area. Two cities may have the same concentration of dust in the air but, if one city's grime is redder or browner than that of the other, the cities may end up with different COH indices.

As a second example, consider an index of forestry. Suppose we calculate the ratio of the number of trees cut down to the number that are planted. When the saws grind all day and few seedlings take the place of the giants, the index is high and environmental quality is correspondingly low. When more than enough seedlings start sprouting, the index is low. This index seems straightforward enough, but there may be areas in a country where many trees could be cut down for years and comparatively little replanting undertaken without much harm to the environment. These regions are where natural reseeding is vigorous. Conversely, there are places where the woodsman would have to spare all the trees and change his trade to seedsman before acceptable environmental quality in the forests would be restored. In these two cases, amalgamation of indices on a national scale would be misleading.

How can these problems be handled? There are at least two approaches. First, there is no reason why only one national value of an index has to be issued. In addition to a national index, we need regional, city, and even local indices. In the case of coefficient of haze, for example, indices for each city could be issued with explanatory notes indicating that they are not directly comparable. Mathematical methods could eventually be worked out to produce comparability.

Second, we could develop indices that include measures of how the environment had previously been treated. In the case of the forestry index, the problem arises because of the crudity of the index chosen, not because the overall concept is faulty. If we knew how much of the forest had been carted away in the past in proportion to the trees left standing, we could revise the mathematics of this proposed forestry index. Then both national and regional indices would have meaning.

There is no inherent obstacle to national indices. Because the ecology is so diverse in most countries, the preparation of a national index requires more thought than does that of local indices. In many cases, the expertise to formulate them is at hand.

Let's look at how some countries have approached the subject. The 1972 edition of the annual report of the Council on Environmental Quality, in the United States, started with a chapter coyly titled "The

Quest for Environmental Indices." A thorough explanation of the scope and limitations of indices was given, as well as short descriptions of the studies that had been funded by the Council, such as the MITRE study.

However, no attempt was made to consolidate the calculated indices into some sort of national value. The Council also suffered from the problem that, although it has a great deal of prestige, it is only an advisory body. The Environmental Protection Agency (EPA), the national environmental action department, has not produced indices as yet.

The CEQ reports tend to be bulky—of the order of 400 pages—and crammed full of numbers and other information. Because of their size and complexity, few except experts can use them to gauge the state of the environment. By way of consolation, the Council and the EPA have under way more studies on the subject. We can conclude that comparatively little progress has been made in the United States toward actually issuing national indices.

Other countries have marched varying distances toward the goal. France has issued what could be a model national set of indices for water, *Inventaire du degré de pollution des eaux superficielles rivières et canaux*. A series of 11 large maps, illustrating the state of French rivers and canals, were prepared. Shown are the values of such parameters as dissolved oxygen, biochemical oxygen demand, and dissolved alkalis, as well as how present levels compare with past values. In addition, a reference booklet was prepared to give experts in the field the method of preparation of the maps.

The information, for almost all practical purposes, is in the form of indices as we have described them. The study tells which French rivers are better than others in terms of water quality indices. The only factor lacking is the calculation of an overall national index so that year-to-year trends can be compared. However, such an index could easily be calculated from the map information. On the basis of this publication, France is ahead of other nations with respect to indices covering as broad a subject as water.

A Canadian study in which I participated employed most available national environmental data to evaluate the state of the country's environment. The index is by no means official. Because of this, it was issued in the journal *Science* rather than as a governmental publication.

The Canadian index divides environmental quality into four areas: air, water, land, and miscellaneous aspects such as pesticides and radioactivity. The air index deals with such aspects as coefficient of haze, sulfur dioxide, visibility at airports, and estimates of emissions of industrial air pollutants. The water index includes turbidity (or cloudiness) of rivers, potentially dangerous metals in drinking water, and mercury in fish. Covered in the land index are topics such as aspects of forestry, erosion, and the availability of national and provincial parks to centers of population. We will discuss many of these topics in the balance of this book. Although the Canadian effort used the judgment of many scientists and environmental administrators, it was still a comparatively small effort.

For the past few years Japan has issued a publication entitled *Quality of the Environment in Japan*. Similar to the CEQ annual reports in the United States, it is perhaps simpler. The Japanese publications contain comparatively little mention of indices as such. However, the levels of many types of pollution are shown, and their official standards are listed as well. As was noted in Chapter 1, if pollution levels and standards come, can indices be far behind? So far, Japan has hesitated to take the final step. Because of the severe environmental problems confronting that country—perhaps the most critical in the world—much effort has gone into measuring pollution and setting standards. Japan has the potential for developing the best set of indices of any nation.

As you can see, the rate of progress among countries toward environmental indices is highly uneven. Some have produced indices on one part of the environment, others are on the brink of producing indices, and still others have unofficial indices. No nation has yet issued a comprehensive and official group, although a combination of the work done so far would be sufficient for the compilation of such a set.

Part of the reason for governmental hesitancy is the natural fear of embarrassment, which would occur if environmental conditions were revealed to be worse than claimed, or if indices had to be recalculated to produce agreement with better knowledge. At least two things are clear about indices: first, they can never be perfect and, second, adjustments will be necessary as we learn more and set different standards. Most governments like to pretend that any information they issue to the voters is the whole truth and nothing but the truth. Environmental indices may be the truth, but no statement or group of statements about our surroundings can be the whole truth. Our environment is too complex for that.

Enough conflicting statements about the environment have been made over the last decade by conservationists and industrialists for the public to realize that opinion on the subject is sharply divided. In consequence, there should hardly be great cries of astonishment if indices have to be revised periodically. When national governments issue environmental indices, they are bound to get a bit of egg on their faces, even if they are good cooks. If the indices are presented for what they are—an attempt to tell the truth, rather than the truth as handed down from above—the egg shouldn't stick.

INTERNATIONAL INDICES

This section should be blank. Politicians, tired of such petty matters as the fact that their best ward captain has the flu or that their campaign posters are misspelled, sometimes like to look at the global picture, and the environment has not escaped their attention. By now we have all read statements saying that the state of the world environment is good, bad, or indifferent. The truth is that we simply don't know what it is. We have bits and pieces of information floating about, but no international organization has yet attempted to fit the data together by means of indices or other devices. The United Nations Environmental Program and other less comprehensive bodies such as the Organization for Economic Cooperation and Development (comprising much of Western Europe, the United States, and Canada) have taken some halting steps in this direction. However, the problems of matching different types of data and standards across international boundaries are immense.

Many of our environmental problems are international, rather than merely national and local. Unfortunately, the day is long off before we can truly grasp the state of these problems by means of indices. As national indices are developed, the probability of international work will become greater, and one day we may have truly global assessment of where we stand.

4

AIR QUALITY INDICES

A monumental statue of a becapped figure holding a sturdy staff stands in a Tokyo plaza. This is hardly a surprising sight; the same model seems to have been the basis for most of the memorials in both the West and the Orient. What makes the scene different from other plazas is the large sign set nearby. The ideograms covering its face are incomprehensible to those of us who haven't studied Japanese, and we might be tempted to dismiss it as an advertisement for something or other. But the sign's right side contains numbers that light up like those on a football scoreboard. From time to time you might see them change. The numbers aren't as innocuous as those recording progress on the gridiron; they tell the concentration of two air pollutants, sulfur dioxide and carbon monoxide. If the flashing sign can be considered as an advertisement for anything, it is for not living in Tokyo.

True, the concentration is not an index. Acceptable standards for sulfur dioxide and carbon monoxide differ significantly in value. Most people don't remember these values, so they can't compare them to the numbers on the sign. As a result, digits that haven't been converted to simple indices probably confuse more people than they enlighten. Nonetheless, the bright lights indicate an overriding concern for one part of the environment—the air we breathe. Figure 8 shows, perhaps a trifle dramatically, that sometimes what we *don't* see is what we get.

WHY AIR INDICES CAME FIRST

Perhaps the most obvious incentive for the production of air indices is our habit of breathing all the time. We can walk away from places where the quality of water or land has been degraded, but we can't turn our lungs off as easily. Air pollution strongly affects two of our five

46

FIGURE 8
By now almost all of us are familiar—too familiar—with the smoke-stacks that loom at the beginning of discussions on air quality. These columns in Birmingham, Alabama, have undoubtedly been photographed to within an inch of their lives. However, how do we know that what we're seeing isn't just harmless steam? We can't con-vict a smokestack on the basis of sight alone. An adequate set of air quality indices would enable us to better understand the effect of these stacks. (EPA—DOCUMERICA, Leroy Woodson, 1972. Courtesy U.S. Environmental Protection Agency.)

senses—vision and smell. In addition, if it's severe enough, we can taste it and touch the residues it leaves behind on buildings and furniture. So far, air pollution doesn't seem to affect our hearing. Any form of pollu-tion that has such an unfavorable effect on so many human senses is bound to have a strong impact on us.

Look at it another way. That highway of scenic beer cans may infur-iate you. That contaminated river may have done in its former finny residents but, if you stay out of it, you probably won't meet their fate. But air pollution can kill people. The events in London, England, and Donora, Pennsylvania, have etched in our minds that air pollution is potentially the most deadly threat to the environment. Because it's a matter of life and breath, we want to know quickly and simply the state of our chances—hence air quality indices.

This is not to say that other environmental problems aren't serious and don't deserve indices of their own. When large numbers of people began to be concerned about the state of the environment, in the 1960s, we reacted in the same way we always have when confronted by a mass of tricky challenges—we tried to analyze the most dangerous one first. As our understanding has grown, we've had the leisure to consider other parts of environmental quality.

WHAT AM I BREATHING, ANYWAY?

The ancients used to think that air was one of the elements, like water or fire. In those days, an *element* was defined as something that couldn't be broken down any farther. A glance through a modern encyclopedia shows that air is made up of more than the two familiar constituents, nitrogen and oxygen. Strange gases like krypton and argon, although in small concentrations, pass through our lungs.

We leave the question of where these odd substances came from to geologists and others concerned with the history of our planet. Humans have adapted to these gases during the course of evolution, so in natural concentrations they're safe. What we're concerned about are the extra added ingredients that our carelessness has mixed into the brew.

Few people can identify all of the substances that we vent into the air, let alone designate their allowed concentrations, a facility which would be necessary before we could calculate a complete air index. We contaminate the air in many more ways than we ever dreamed. It's not our object here to list every air pollutant and its effect on the human body, since surveys of this type have already been done. If they were combined we'd have a gargantuan volume the size of the MITRE report mentioned in the last chapter. Basically, we can divide air pollutants into two broad categories—the masses and the gases.

I'm calling a mass everything that can't be classified as a gaseous compound. Masses include substances like asbestos, lead, and what is called total suspended particulate matter. This latter material is usually known as smoke, dust, or grit.

When we open the cupboard of gaseous compounds that we've added to the air, we see that it's the opposite of Mother Hubbard's. The shelves bulge with gases such as hydrocarbons, oxidants, and carbon monoxide. Every so often, the stockboy adds a new compound such as vinyl chloride. Can we devise understandable air indices for this hodgepodge of pollutants?

We can—if we keep in mind the relationships between concentration levels, the times during which these levels occur, and standards for each pollutant. This mental juggling may seem tricky at first, but it's straightforward if we focus our attention on one pollutant at a time. The next section explores some of the aspects of this juggling.

HOW STANDARD IS A STANDARD?

The title of this section may at first sight seem nonsensical. Surely a standard is, by definition, unchanging and applicable to all situations. However, air pollution is such a complex subject that we have to adopt different types of standards for different conditions.

All of this may sound vague. Let's clear it up with some examples. The grime we call total suspended particulate matter comes from more than one source. Some of it may come from oil burning, some from industrial processes, and some from natural causes such as dust storms. The particles are a mixture of different materials and sizes.

Health effects of dust depend on the size of the particles. The exact details of the research showing this aren't important here. What does matter is that, to devise an index of particulate matter, we may need more than one standard, depending on the size of the dust particles. Specifically, we may eventually have standards for dust particles less than 2 microns in size, between 2 and 5 microns, between 5 and 8 microns, and so on. (A micron is a millionth of a meter, or about 40 millionths of an inch.) The concentration of dust in each category depends on the process by which it got into the air in the first place. Instead of one standard for one pollutant, we may have two, three, or more.

The same problem occurs when we consider how pollutant concentrations change with time. A low average concentration may include some periods during which the concentration is high. For example, suppose that we had a city whose average concentration of a certain air pollutant over 1 year was 1 part per million (often abbreviated as ppm). If the standard for that pollutant were 2 ppm, and we calculated the index in the simple linear way we outlined in chapter 1, the average index for the year would be ½, or 0.5.

Let's now further suppose that one not-so-fine day pollution control devices all over the city break down. Catalytic converters tumble off cars, electrostatic dust cleaners fall down smokestacks, and fumes of all sorts fill the beleaguered city. The concentration for the pollutant we're interested in goes up to the unprecedented level of 37 ppm. Fortunately, enough string, baling wire, and welding shops are available to enable us to reinstall the pollution control devices, and the next day the concentration falls to the usual level of 1 ppm.

The effect of that one day's concentration on human health is probably more severe than that of the rest of the year put together. Yet the average index for the year would only increase by 10% as a result of this episode. (The arithmetic is as follows: there are 364 days with an index of 0.5, and one with an index of 37/2, or 18.5. The total number of "index-days" is then $364 \times 0.5 + 1 \times 18.5 = 200.5$. The average for the 365 days in the year is then 200.5/365, or 0.55, that is, 10% more than the original value.) We have to devise a method of calculation that will give greater emphasis to the hours, days, or weeks with elevated concentrations. If we can't, high values of an index will tend to be masked by comparatively low ones.

There are an unlimited number of ways for preventing this swamping effect. One approach is described in the MITRE report, referred to above. They employed the Kronecker delta, so called because the Greek letter delta is used to represent it. To illuminate this rather obscure corner of algebra, an analogy would be your decision whether or not to put on your gloves when you go out tonight. If the temperature is 50° or below, you put them on; at a higher temperature you go gloveless. You've been using a Kronecker delta, with a temperature level of 50°, without even realizing it. Without getting into the mathematics, the delta can be described as having a value of 1 when the quantity it deals with is above a specified level, and 0 when it is below that level. For example, if we set the Kronecker level at a value of 0.5, the delta would equal 1 if indices were 0.5 or above, and 0 if they were below.

The concept may be made clearer by considering another hypothetical example. Suppose that we had three daily air indices of values 0.1, 0.2, and 0.3. Let the Kronecker delta cutoff point be 0.5. The average of the three values is 0.2. Each of the values is below 0.5, so when the delta is used on them they are each multiplied by 0. The operation with the deltas then adds nothing to the average of 0.2.

Now suppose that a fourth value of 0.7 is added to the group. The average of the indices is now 1.3/4, or 0.35. When the deltas are applied to the four values, the first three are multiplied by 0, but the fourth is multiplied by 1 since it exceeds the standard of 0.5. We then add 0.7 to 1.3, obtaining a total of 2.0. The average is then 2.0/4, or 0.5. In this case, values above 0.5 are counted twice in the average. The Kronecker delta is thus a way of giving extra weight to values that exceed a predetermined level.

Another general method for emphasizing the more extreme values of a collection of numbers is the root mean square operation. The squares of numbers always differ more from each other than the numbers themselves do. In the root mean square procedure, the numbers are squared, their average is taken, and the square root of the average is calculated. In the example above, the average of the first three indices 0.1, 0.2, and 0.3, was 0.2. If we take the root mean square approach, we square each of the indices and then add them, to obtain $0.01 + 0.04 + 0.09$, or 0.14. The average is 0.14/3, or 0.0467. The square root of this quantity is then 0.216, about 8% larger than the simple linear average. The root mean square calculation was used in both the MITRE report and the Canadian study.

Working with these mathematical tools shouldn't make us forget that the most important quantities are not the tools themselves, but the results the tools carve out—the environmental indices. At the moment, we don't have enough measurements (or, in the parlance, an adequate statistical base) to decide which of the many possible tools should be used. More data will allow us to resolve this question.

A standard isn't as unchanging as it sounds. Depending on the type of pollutant, we may have to use more than one standard in calculating an overall index. For example, the MITRE index used different sulfur dioxide standards for a year, a day, and a 3-hour period. Its complicated calculations were designed to make real health effects show up mathematically. From this point of view, using more than one type of standard is just a way of improving the relation of an index to reality.

CHANGING STANDARDS

As if having more than one standard for a given pollutant weren't enough, these standards can change. There are two main reasons for their alteration: research and politics.

Environmental studies can show that a standard or set of standards is either too high or too low. There are at least two reasons why we shouldn't be too surprised if new results are brought to light. First, much of the work on which present standards are based is not as thorough as it could be. Much of it was done in the days when the word *environment* meant how you were brought up by your family, not the state of your natural surroundings. As our knowledge increases, changes in standards are bound to occur. Second, a large part of environmental research on standards is concerned with the long-term effects of pollutants, and this work naturally takes a long time. It may be comparatively easy to set a standard of x ppm for pollutant Y for one day's exposure; setting a limit for a year takes subtle research that may last a year of Sundays.

Standards may also be partially based on political considerations. For example, an objective for a given air pollutant may be z ppm, set after consultation with scientists. If concentrations of this pollutant rise above this value, i.e., the index is greater than 1.0, legal regulations are likely to close down industries and restrict other activities, such as motoring, that produce this pollutant. These actions naturally lead to pressure on governments to relax the standards by increasing them. This makes the standards harder to exceed, and the computed index of this pollutant may seem to be getting lower; in other words, better. To make a fraction smaller, you must either lower the numerator (reduce the concentration) or increase the denominator (raise the standard). Just as in the old chestnut about alleviating flood conditions by either lowering the river or raising the drawbridge, the latter is easier. Hence the temptation for politicians.

It may be galling to us that scientists have their carefully constructed standards sacrificed on the altar of political expediency. Under our system of government, elected representatives have the final say. As the public gradually gains more confidence in scientific standards, the effects of political pressure will probably diminish.

A particular case of changing standards in the opposite direction may be seen in Japan, where, according to a 1972 document, the standard for sulfur dioxide for "excessively populated areas" was 0.38 ppm from 1967 to 1973, 0.27 ppm for 1973 to 1978, and 0.20 ppm after 1978. As the standard becomes lower, the index will increase if the concentration of sulfur dioxide remains the same. The apparent political decision here was to make the standard relatively high for the first few years of the period under consideration so that any index computed during that time would be relatively low.

Does fiddling with the standards produce off-tune indices? Not any more than it does economic and social indices. Chapters 1 and 2 briefly discussed this problem. Standards for economic indices can change and have changed without ruining the fundamental concepts. For example, in computing the consumer price index the average market basket of goods and services bought by consumers is altered in most countries every 5 or 10 years, as preferences change. In calculating the index of unemployment, modifying the question from "Are you out of work?" to "Are you looking for a job?" has strongly changed the value of the index. These indices are not made useless by changing standards; we simply have to remember the quantity to which we are comparing our numbers.

Even if politicians went on permanent vacation, environmental standards would change as we learned more about the effect of pollutants on our health and belongings, and as we demanded higher levels of such amenities as parkland. As standards change, we may have to recompute historical values of the indices. The additional arithmetic will be worth the increased understanding of how the state of our environment has shifted.

LINEAR OR CURVED?

Dividing the concentration of a pollutant by the standard to obtain an index may seem like the most natural arithmetical operation available. But when we do this, we're making a hidden assumption. We're implying that the effect of the pollutant, which the index is an attempt to measure, is *linear* with concentration. As an example, for every doubling of the concentration the adverse effect of the pollutant doubles.

This may not always be true. Some scientists claim that the effect on health of pollutants is *nonlinear* and that the effect goes up faster than the concentration. For example, a doubling of concentration might triple the effect on health; therefore an index for this pollutant should also go up by a factor of 3, rather than 2, as we would ordinarily expect. It's true that sometimes the human body reacts in a nonlinear manner. A small increase in a drug dose may trigger a violent reaction that is unexpected on the basis of slightly lower doses.

If environmental effects on health were generally nonlinear, the concept of indices would still work. Instead of using a simple fraction to compute them, however, we'd have to determine indices from a curve on a graph of index versus concentration. We haven't used the concept of nonlinearity here because nobody has proved conclusively that the effect is there, although it is suspected for certain pollutants. Eventually scientists may find the shape of the elusive curve to their and our satisfaction, but regulating agencies now usually assume linearity. In the back of our minds, though, should lurk the thought of possible nonlinearity.

SCALES—UP, DOWN, AND SIDEWAYS

It's nothing short of amazing how many assumptions can be packed into something as seemingly simple as a numerator sitting on top of a denominator. Consider the scaling used. We have implicitly assumed that, when the index is 0 (a concentration of 0), the part of the environment to which it refers is "perfect." When the concentration equals the standard, the index is 1.0, which indicates a more-or-less serious situation. If the concentration of the pollutant increases, the value of the index will be greater than 1.0. The scale then runs from 0, for perfection, to a large number, for environmental disaster.

In principle, a scale of indices could range from any number, positive or negative, to any other number. We've chosen the present scale primarily for its simplicity, but it's easy to think of others. For example, the consumer price index compares a measured quantity to a standard, just as we have done, but the resulting figure is multiplied by 100.

There are an endless number of possible scales for environmental indices, but two should illustrate them. First, we can divide the standard by the concentration, instead of the other way around. By using the scale we described in the last chapter, it might be claimed that we do not have an index of environmental quality, but rather one of environmental pollution. If this is true, the scale of the index should run from a very large number, when the concentration of a pollutant is low, to 0, when the concentration is high. The two conditions would then correspond to high and low environmental quality, respectively. There are many ways to produce this mathematically, but probably the simplest is to invert the fraction we have been using so far.

A second type of scale that has been suggested is the so-called bidirectional index. This is based on the assumption that, when the concentration of the pollutant equals the standard, the index of environmental quality should be zero. Again, there are many mathematical methods of achieving this result, but a simple one subtracts the original index we have been using (concentration/standard) from 1. When the concentration is 0 (a "perfect" condition), the index is $1 - 0 = 1$. When the concentration equals the standard, the index is $1 - 1 = 0$. If the concentration is greater than the standard, the quantity to the immediate left of the equality sign becomes greater than 1, and the index becomes negative.

These are only two of the many ways scales can be constructed for indices. Unfortunately, there is no infallible method, handed down from the heavens, that is guaranteed to eliminate any misconceptions. Every scale has its drawbacks. The one used here, concentration/standard, is discussed in detail because many scientists have employed it and because it is simple.

As environmental indices come into wider use, some scales may be governed more and more by public opinion. For example, it's unlikely that the man in the street will ever have the statistical knowledge required to set scales based on health effects of asbestos in the air. However, he might take part in the developing of scales based on the appearance of water and the availability of parkland, as these are based on personal preferences. Because one of the prime purposes of these indices is to inform everyone of the condition of his surroundings, the scaling should be done so as to convey the message clearly. In the long run, everyone must add his or her voice to the discussion.

SYNERGY

Not long ago almost every magazine institutional advertisement for up-and-coming companies had the word *synergy* trumpeting from the page. "Look at us," they would cry, "we have hardly any assets, our balance sheet is in glowing crimson, but we've got synergy." The flourishing of the word was somehow supposed to make the reader look more favorably on the company. It rarely did, since most people either never quite got its meaning or passed it off as another advertising gimmick.

This lack of understanding was unfortunate, and not merely from the point of view of Madison Avenue. *Synergy* does actually mean something. It signifies that sometimes the overall effect of two or more actions is greater than just the sum of the actions. In other words, in a synergistic situation the whole is not equal to the sum of the parts—it's greater. For example, if you open only your left eye, you will see a flat two-dimensional scene. If you do the same with the right eye, you will see the same scene from a slightly different angle. When you use both eyes, you see a three-dimensional scene after your brain superimposes the two pictures. The three-dimensional view could not have been anticipated on the basis of the two one-eyed views, hence the overall effect is greater than the sum of its components. Synergy is on the march.

How is this concept related to environmental indices? Some environmentalists say that the combination of two pollutants at the same time can be more dangerous to humans than the sum of each of them taken separately. A well-known medical example is the more than doubly dangerous effect of barbiturates and alcohol. If synergism does apply to pollutants, it should be reflected in the index. For example, suppose that the index of sulfur dioxide is 0.2 and the index of suspended particulate matter is 0.3. There are, of course, many complicated ways of incorporating indices the two into one.

For simplicity's sake, let's assume that we can combine the two indices into an overall air quality index by averaging them. Then the value of this index is $(0.2 + 0.3)/2 = 0.25$. Now suppose that there is a synergistic effect between the two pollutants. The combined index should then be higher than 0.25; perhaps 0.3 or 0.4.

While it is highly likely that synergisms exist among and between pollutants, few calculated indices have taken account of them. It has proved difficult to calculate the precise extent of the synergism. Without an exact knowledge of the effect, we cannot include synergism in our calculations.

This emphasis on definite values is both a weakness and a strength of environmental indices. It is a weakness because situations that are not completely well defined have to be left out of the arithmetic. It is a strength because it forces us to put all the cards of our assumptions on the table, even though we'd rather that some of them were kept up our sleeve.

To a limited extent, synergism is included in the Ontario air quality index, where the effect is considered for two pollutants. It is possible that, as we gather more knowledge about the combined effects of pollutants, we shall be able to incorporate synergism into other indices.

TWO APPROACHES TO AIR QUALITY INDICES

There are two main paths we can follow in computing air quality and other environmental indices. First, we can *measure* what is in the air, water, or other medium into which we drop our wastes. Second, we can *estimate* how much pollutant we are putting into the air or water.

The difference between the two methods seems at first to be more a matter of air-splitting than reality. After all, if we can estimate how much smoke, soot and ash we loft into the skies, we should have a good idea of what is actually there. Part of what goes up must not come down—the atmosphere can't have its dirt and consume it.

The key difference between measurements and emission estimates lies in how pollutants are distributed in the air or water. For example, particles of lead, emitted from car exhausts, are important pollutants. Suppose that we make an estimate of the amount of lead entering the atmosphere each year. This could be done by multiplying the average amount put out by each car by the number of cars, if we assume that there were no other sources.

Now suppose that the number of cars in your town increases by 10%. If we assume that the amount of lead per car remains the same, the volume of lead rises by 10%. An index of estimated lead would also rise by this proportion.

You don't live all over town; you live in one area. It may be that most of the increase in driving is near your block. In this case, the measured lead index for your part of town might increase by a large factor, while the indices for other parts remain the same. When averaged over the whole town, the measured index could well rise by the 10% expected on the basis of the emissions estimate. But you and your neighbors wouldn't be satisfied with that result if you suffered from lead poisoning as a result of the increased traffic.

This example illustrates the problems that arise when indices are computed from estimates. Basically these indices are too general. What they don't take into account is that pollutants distribute themselves non-uniformly over the land- and waterscape. Weather conditions are among the main reasons why air pollutants tend to concentrate in some areas and avoid others. Because these pollutants are gases and light particles, it takes only a slight breeze to move them a long distance.

The problem is analogous for water pollution. This question is discussed in more detail in the next chapter. Emissions-based water indices would be perfectly satisfactory if, at every place where wastes are discharged into rivers, somebody stood with a giant Mixmaster, making sure that water and wastes were perfectly churned up. Nobody is going to do that, so parts of some rivers and lakes have a much higher concentration of these wastes than others.

If emission indices have such drawbacks, why are they used at all? Primarily because they are a cheap and convenient method for making rough estimates of environmental quality when no other methods are available. Chapter 1 noted that we can't have instruments measuring pollution at every street corner, if only because of the cost and personnel required. Where actual physical measurements aren't available, estimates of emissions are the "next best thing to being there." This shortcut works fairly well when only small areas are considered. However, when we try to derive national or global air indices from estimates, the ice on which we're skating becomes drastically thinner. The variations from region to region become so great that the concept breaks down. Therefore, if we do compute national air indices, they should be based primarily on measurements. Estimates can be used to fill in any information gaps.

MEASURED AIR QUALITY INDICES

Trying to describe who measures what, when, and how—and the results of all of this activity—would rapidly produce an encyclopedia, not a book. We only touch briefly on the pollutants that are likely candidates for air quality indices.

Sulfur dioxide and particulate matter, both generated primarily from industrial activities, have traditionally been the two main starting points for these indices. In fact, the Ontario air quality index, mentioned in Chapter 3, uses only these two. We don't dwell too long on the standards used for these indices, since they have varied somewhat with time and place. However, some recent standards for sulfur dioxide have been 0.02 ppm on an annual basis, 0.10 ppm on a daily basis, and 0.50 ppm on a 3-hour basis. This pollutant has not one, not two, but three standards to make the mathematicians happy. In effect, our lungs are allowed to have a higher concentration for short periods of time.

A typical annual standard for particulate matter is 60 micrograms per cubic meter of air. A microgram is a millionth of a gram, or about 35 billionths of an ounce. To put these units back into the parts per million notation with which we're at least a little familiar, 60 micrograms are roughly equivalent to 0.00003 ppm. The concentration appears to be much lower than that of the previous pollutant because in this case we're considering solid matter, which is much denser than gases like sulfur dioxide. Because particulate matter varies in composition and density from one location to the next, the unit of the standard is usually in micrograms per cubic meter of air. The standard is 150 micrograms per cubic meter on a daily basis, or about 0.000075 ppm. Again, the amount we're allowed for a short time is much higher than the long-term amount.

In addition to sulfur dioxide, other compounds of sulfur and oxygen are produced during combustion. Occasionally the collection is referred to as *sulfur oxides,* and in principle an index could be devised for the group rather than for sulfur dioxide alone.

The term *coefficient of haze* is sometimes used in place of particulate matter. The pollutant measured is still the same—dust and other particles in the air—but a somewhat different method is used. Again, we can construct an index for coefficient of haze by using the same sort of reasoning as used for particulate matter.

Oxides of nitrogen is a collective term referring to nitrogen oxide, nitrogen dioxide, as well as other compounds of oxygen and nitrogen, in analogy to sulfur oxides. Most measurements have been done on nitrogen dioxide. Typical standards have been 0.05 ppm (in the U.S., on an annual basis) and 0.5 ppm (in Japan, on an hourly basis).

Most carbon monoxide is emitted from automobile exhausts. Standards have been set for 1 year (5 ppm, in Japan), 8 hours (9 ppm, in the United States), and 1 hour (35 ppm, in the United States).

The multiplicity of standards for these pollutants may prove confusing at first. An analogy might be an ice-cream cone standard. You might be able to eat 3 cones in an hour without a stomachache (multiply this standard by 2 for those under 10 years), and about 15 in a day, or about 1 per waking hour. A devotee of ice cream might have 500 cones in a year, or about 0.1 per waking hour. The standards for ice cream and pollutants decrease as the time being considered becomes greater.

Two types of pollutants that produce photochemical smog similar to that in Los Angeles are oxidants and hydrocarbons. High and brief short-term concentrations of oxidants are more dangerous than those low and long-term, and a typical standard is 0.08 ppm on an hourly basis.

Many pollutants other than those already mentioned can serve as the basis of indices. As we change from one fuel or type of combustion to others, the chances are that new sorts of pollutants will find their way into the air. For example, the presence of lead in the air was unknown before lead was added to gasoline. Rather than list every possible pollutant here, which couldn't be done in any case, we only point out that any overall index of air quality must allow room for the addition of new hazards to health.

ESTIMATED EMISSION INDICES

As we mentioned previously, estimates of air pollution can be used in the calculation of indices. An example of their complicated nature is found in the Manchester study, which was cited in Chapter 3. Estimates of traffic density in various parts of the city were made. To calculate the resulting amounts of pollutants, the proportion of auto and diesel traffic had to be estimated, since internal combustion and diesel engines produce different quantities of pollutants. A determination was then made of the average gasoline mileage for the two engine types. The final estimation was the quantity of each pollutant created per gallon of fuel. The four factors were then multiplied together to produce a grand estimate of the gross amount of pollutants.

This method suffers from at least two drawbacks. First, estimates are piled on top of estimates until the whole structure threatens to collapse under its weight. Second, there are no real standards against which the estimates can be compared. For example, in the Manchester study it was concluded that the amount of carbon monoxide emitted in the central part of town during a certain period of time was about 3000 pounds per square mile (or about 490 kilograms per square kilometer). But this is just a number on a piece of paper; we have no standard with which to compare it.

The problem was handled, not solved, in the Canadian index mentioned in Chapter 3, by comparing the estimates for particular areas to the estimated average for the whole country. For example, if the calculated amount of pollutant per person in an area is 2 units and the national average is 4 units, then the index for this quantity is 2/4, or 0.5. This is not a perfect approach to calculating indices, but it can sometimes be the only possible one.

Emission data can be calculated for particular industries as well as for sources like automobiles. For example, an estimate of pollution from the steel industry might take into account that $X\%$ of steel mills have pollution-control device Y, $Z\%$ have device Q, and so on. From this information, we can estimate the total amount of pollutants per ton of steel produced. How do we calculate an index from this? One way would be to estimate how much of a particular pollutant an industry would produce if it employed the best available environmental technology. Then the index would be the actual amount of a pollutant divided by the lowest amount theoretically possible. For example, if the quantity of pollutants being produced weighed A tons and the lowest amount using advanced technology was T tons, the index would be A/T. One problem this index would have is that it could never sink below 1. To rectify this problem, it may be ncessary to use some of the subtler forms of mathematics mentioned elsewhere in this chapter.

CAN THE TWO TYPES OF INDEX BE COMBINED?

We seem to be confronted with the problem of trying to add apples and oranges. More accurate indices are produced when actual measurements are used for their calculation. On the other hand, estimates of emissions are easier to obtain, since they can often be produced without spending much money on instruments and highly paid technicians. If we keep each type of index separate, no problems arise. However, if we want to calculate regional and national air quality indices, we need some way of combining the two. How can we handle the emission indices so that double-counting is reduced to a minimum?

Inadvertently counting the same thing twice or even three times is a problem that all index-makers face. The economists who calculate the gross national product count the value of bread, but are careful not to add the value of the dough from which the bread is made. Adding the value of the dough to that of the bread would be counting some quantities twice.

We face the same problem in combining measurement and emission indices. For example, consider the emission index for the steel industry, mentioned above. Part of the pollution from these factories would also be detected by instruments and show up in the measurement indices. Some of the pollutants would then be counted twice.

One solution to this problem, adopted in the Canadian index mentioned in Chapter 3, is to use emission indices only where measurement indices aren't available. For example, the emission type of index was available for all industrial sources of sulfur dioxide. In some cities this pollutant was being physically measured. The emission index for these cities was then not used in the calculation of an overall national air index. The only part of the emission index used was that related to areas for which there were no measurements—generally smaller cities and rural areas. In this way, double counting was reduced to a minimum.

This procedure may seem like yet another layer on the cake of mathematical abstraction. However, if indices are eventually to be used by the public and government policymakers, we have to make sure that they are as accurate and fair as possible. Sometimes it takes a heap of calculations to make a measurement and an estimate into an index.

SEEING IS BELIEVING

There are other, less well-known ways to determine the state of the air. One method of calculating a simple air index was developed about 100 years ago. An American scientist named Vider in Tokyo observed Mount Fuji, a sacred mountain about 60 miles away, in 1877 and 1878. On the average, he could see it about 1 day in 3, although the visibility depended on the weather conditions, which in turn depended on the time of year. A comparative study was made in 1971, when the mountain could be seen only about 1 day in 7 on the average. Here we have the makings of an uncomplicated air index—visibility.

Visibility is an average of the effect of a number of different pollutants, and an index based on it is not quite comparable to the others mentioned in this chapter. For example, particulate matter in the air will certainly cut down on how far we can see, but some of the gases we've discussed will do the same under certain conditions.

Can we make an index out of visibility readings? To do so, we must have both data and a standard or standards. Many sources of data are available. Airports throughout the world make visibility readings many times a day, to help in aircraft navigation. If anything, we have too many numbers rather than too few. The measurements suffer from being taken at airports, not in cities. As anyone who has ever blanched at an inflated taxi fare out there knows, airports are frequently a long distance from the center of town. However, it's been shown that airports near highly polluted cities generally have low visibility, so there is a strong relationship between indices computed for airports and for nearby cities.

One problem with visibility indices is that there are no available standards. It appears unlikely that governments will set standards in the near future. A way out of this impasse is to use the visibility at airports known to be far away from major sources of pollution as a standard. There are still some left.

In an index of visibility, low visibility corresponds to high levels of pollution. At the beginning of this chapter, we adopted the convention that a low index would indicate low pollution levels, or high environmental quality, although a case could be made for having the scales running in the opposite direction. Continuing with this convention, a visibility index would have to be upside down compared to the indices of the other pollutants we've discussed. The standard for visibility would then be divided by the measured quantity (the visibility at "nonstandard" airports). If the measured visibility were low, the index would be high, and vice versa. More mathematical finagling could make the index go to 0 for the best possible visibility, but we won't go into dull details here.

A SMELL BY ANY OTHER NAME

Air pollutants create many health hazards and kill plants, but most don't immediately drive us away from the place where they're emitted. One type does—smell. Because the human nose is so sensitive to even tiny concentrations of certain foul odors (in the range of parts per billion rather than the parts per million we've been talking about so far), these types of pollutants might seem naturals for producing indices.

In spite of this, no comprehensive odor indices have yet been devised. The machine has yet to be built that can tell the difference between various types of smell and their concentrations at the same time. There is no inherent reason for preferring an instrument to a human if people can do the job; after all, visibility readings are taken by eye. Unfortunately, the nose has even more trouble than machines in distinguishing between smells and their levels. We can tell that something is wrong in the air, but we often can't tell what it is or how strong it smells.

Although an index of smell would be a significant part of an air quality index, information now exists primarily in the form of scattered studies. As these are put together to yield comprehensive understanding, we shall be able, so to speak, to smell our way to an index of odor. The first sniffs towards that goal are taken in Chapter 8.

ACHTUNG! ACHTUNG!

Rather than just computing averages for use in indices, another method of evaluating pollutant concentrations is by adding up how often short-term standards are exceeded. This type of index could be called a *warning* or *extreme value* index. The former name applies because many governments issue instructions to industries to lower or even cease their production when standards are exceeded. For example, if 10 warnings are issued out of 250 working days in a year, the index is then 10/250, or 0.04. The MITRE study considered an index of extreme values, employing Kronecker deltas and other sophisticated mathematics. These types of indices have the advantage of focusing attention on the worst incidents of pollution, but they usually ignore levels of concentration that, although high, do not exceed standards.

OTHER TYPES OF AIR INDICES

Many other types of air quality indices can be devised, or at least thought of. For many of these, measurements are spotty and standards don't exist. However, as we learn more about our environment, we may be able to use some of them. Possible candidates might be corrosion damage by pollutants; agricultural crop loss; the number of regions, states, or provinces with adequate air pollution standards; the amount of solar radiation (a measure of how much of the sun's rays we're blocking off); and the number of plants that are damaged by pollutants. Some particularly sensitive types of vegetation are tobacco and orchids. We discuss these last indicators in Chapter 7.

SOME TYPICAL AIR INDICES

The MITRE study is probably one of the most comprehensive attempts so far at air index formulation. A total of five air pollutants were covered, and these were combined by a root mean square method. Each of the indices takes account of standards for differing time periods by means of Kronecker deltas and other devices. In addition, an extreme value index was calculated to consider the highest values of four air pollutants. Finally, a list of other potential air indices was prepared. Since data do not exist for many of them, the indices calculated for this last-named group are theoretical.

The Manchester study considered a number of parameters, including grit, dust, and visibility. Emission indices were also calculated. Only smoke and sulfur dioxide were used in the final set of measured indices. A simple linear scale was used in which the highest concentrations of these pollutants in the Manchester region were assigned a value of 100, the lowest a value of 0; values for intermediate concentrations were interpolated between these numbers. For example, if the highest concentration of a pollutant in the region were 220 ppm and the lowest 20 ppm, a reading of 120 ppm would produce an index of 50, since it is halfway between the ends of the scale. The two measured indices were simply added together, without root mean square or other involved calculations.

A study in Oak Ridge, Tennessee, used five of the major pollutants, as did the MITRE work, but somewhat more complicated mathematics was employed. The statistical treatment, which was too involved to describe here, is discussed in detail in the Thomas book. One objective was to calculate weights for different pollutants so that each was given its appropriate significance in the devising of an overall air index.

The National Wildlife Federation air index, also described in

Thomas, was based on human judgment and used no mathematics.

Finally, the Canadian air index is divided into three parts. The first used the root mean square method to deal with measured pollutants, in a manner similar to that of the MITRE study. The second produces a visibility index at many Canadian airports. The third is the calculation of an emission index for sulfur dioxide and particulate matter for areas where there were no physical measurements. The three parts were combined into a national air quality index by means of mathematical weights supplied by experts in the field.

CONCLUSIONS

Because of the many measurements that have been made of air quality, these indices have been the first to appear. Indices relating to other aspects of the environment haven't yet attained the same degree of mathematical maturity and understanding. As control measures based on these air quality indices begin to be applied, we may be able to say again, as Shakespeare did in *Macbeth* almost four centuries ago,". . . the air/Nimbly and sweetly recommends itself/unto our gentle senses."

5

WATER INDICES

In modern times, we can transform an old saying into "Water, water, everywhere, but not a drop to drink, or bathe in, or boat on." Conditions aren't as bad as this in most places, of course, but degradation of water is one of the most noticeable forms of pollution. We can use indices to gauge the extent of the deterioration in much the same way that we did for air. In this procedure, we have to keep in mind how water differs from air.

DIFFERENT USES—DIFFERENT ABUSES

We breathe for only one purpose—to sustain life. We don't breathe one kind of air for recreational purposes and another for supplying oxygen to our bodies. As a result, we are really dealing with just one type of air index although, as the last chapter shows, there are many ways of approaching the subject.

When we calculate indices of water quality, the situation is different. We have varied uses of water, as shown in Figure 9, so we have varied indices. Let's take a few examples. When we consider drinking water, we have to be concerned about chemicals and other pollutants in the parts per million range of concentration. We can then set up indices that compare their concentrations with those in approved standards.

Suppose that you then go for a swim in a nearby lake or river. Its concentration of pollutants may be higher than that considered acceptable for drinking water, but you suffer less potential damage to your health unless you swim with your mouth open. For water used for swimming the allowable concentrations are much higher; in other words, the denominator in the usual fraction for an index is greater. When this occurs, the value of the index becomes smaller. We really

FIGURE 9
The uses of water can vary from day to day or hour to hour. Water quality indices must take these changes into account. This smelt run in Michigan could be used as a beach tomorrow and a haven for sailboats a week from now. Evaluating these uses in terms of desirable water quality can be done, but it's no simple task. (EPA—DOCUMERICA, Donald Emmerich, 1973. Courtesy U.S. Environmental Protection Agency.)

have two indices—one for the water we drink and one for the water we swim in.

If you get tired of swimming and decide to go boating instead, we have to devise different indices for this water use. You probably won't be as concerned with relatively small concentrations of pollutants in the water as with the areas covered by algae and obstacles, such as logs and debris. We can then devise a third group of water indices, to deal with these environmental problems.

Or put yourself in the position of a fish. Our finny friends are known to have different sensitivities to water pollutants than humans do. For example, trout may become diseased or die when the concentration of pollutant X is y ppm; no effect may be noted in humans until the level is z ppm. The standards for fish are often different from those that apply to us. We then need a fourth set of indices, to apply to water inhabitants.

The need for more than one set of water indices may complicate the situation, but shouldn't confuse it. Some parts of our rivers, lakes, and seashore have more than one use, and we must carefully interpret the various indices that are calculated. Because many of us are already adept at this interpretation, the problem isn't completely new. For example, weather bureaus often predict expected monthly temperatures and rainfall. In effect, these forecasts are indices, since they compare what they claim will happen (the value) to the usual conditions prevailing at that time of year (the standard). Farmers plant many crops, ranging from corn to wheat to oats. Each bases his sowing and fertilizing schedule on that aspect of the weather index of interest to him and disregards the other aspects. If he plants corn, which needs wet weather to germinate, he may irrigate when the weather index predicts relatively dry times ahead.

In this and many other ways we have learned to pick and choose among a wide variety of indices. Water indices can be used with the same degree of differentiation.

A SOLUTION TO POLLUTION IS DILUTION

Suppose that a company, following its usual practice, dumps its daily dose of sludge and waste chemicals downstream from its plant. But this time the situation is different: an alert regulatory agency has installed a monitoring instrument a few yards away from the source of pollution, on the messy side. The readings show that a standard for fish or human health is being exceeded; that is, an index of water quality is relatively high.

Hauled before a court or the bar of public opinion, the company defends itself. "It's true that the index is high not too far away from our outlet pipe. But let's be reasonable. All the fish in the river aren't congregating at the point where the measuring was done. They're spread out all along the stream. So, for that matter, is our effluent. The water dilutes it farther down, so that the concentration of pollutant is low there, as is the index."

This, in a slightly enlarged nutshell, is the reasoning behind the cryptic slogan, "A solution to pollution is dilution." Water indices improve, in general, as more mixing occurs or as the rate of river flow and streams increases. Even if Noah's neighbors had all been terrible polluters, water quality indices would have been very low, had they existed at the time of the Ark, because of the tremendous volumes of water floating it.

Chapter 1 noted that part of any environmental index is attributable to "background," or natural causes. Water indices are affected by the flow rate of rivers, which, in turn, is altered by the amount of rain that falls. As we consider these indices, we have to make sure that dilution effects don't confuse our understanding of the state of the water.

TO PREPARE, JUST ADD WATER: PARAMETERS AND POLLUTANTS

When we considered air indices, life was comparatively simple. We decided that whatever mankind carelessly spews into the air is a potential pollutant, and that was that.

The situation is not quite as simple with indices of water quality. We can alter the water without adding something as well-defined as a pollutant. These quantities, which we can call *parameters,* are still a cause of environmental degradation, though not in the same sense as a particular chemical or type of particle.

For example, consider heat. We all know of cases of power plants that use river water as a coolant. Nearby aquatic life often can't take the heat in the aquatic kitchen, and is killed or driven away. We can devise an index based on the rise in temperature, in which the standard is the increase in temperature that doesn't significantly harm fish and plant life, and the value of the environmental quantity is the actual temperature rise. As usual, the index is the value divided by the standard. This index is a useful guide to events around a power plant, even though the temperature of water isn't a water pollutant as we usually understand it, but rather a water parameter. We can use both in devising water indices.

A parameter in one environmental system isn't necessarily one in another. For example, the temperature of the air isn't now regarded as an environmental parameter. If it were, Dallas and Miami would have extremely high air quality indices and low environmental quality. It's entirely possible that we've been increasing the average overall temperature of the earth through our industrialization. We haven't yet been able to devise an index of this effect that will adequately separate out natural fluctuations.

In addition to temperature, other parameters that we shall consider in water indices are biochemical oxygen demand (BOD) and dissolved oxygen (DO). The former is a measure of how much oxygen is required to decompose organic matter in water, and the latter a measure of how much is there. The levels of both are altered by the addition of pollutants to water, although neither is a pollutant in itself.

Chapter 1 pointed out that an environmental quality index isn't solely a pollution index. Dealing with indices based on water parameters as well as pollutants makes that claim more understandable. Subsequent chapters continually show that environmental indices are more than the sum of indices of pollution.

MEASUREMENTS AND EMISSIONS

The battle between using actual measurements to determine environmental indices and employing estimates of effluents for the same purposes, begun over air indices, here enters its second phase. We simply don't have enough instruments to produce all the measurements needed for complete water indices, so we frequently have to rely on estimates. For example, consider the case of dissolved oxygen. The DO "lag," or place where its value drops significantly, is usually near a source of pollution. If we know its location and position our instrument near it, we can use the size of the lag to calculate a DO index. The place where the lag occurs often moves around, due to changes in the amount of pollution and in the flow rates of rivers. We can't keep racing back and forth with our sampling instrument, so we often depend on estimates of the oxygen deficit that is created by the pollution source.

Let's take another example of measurements versus estimates. In the Canadian index referred to in Chapter 3, indices were devised for the level of wastes from municipal sewage plants. We would have a better handle on the truth if measurements were made of the quality of water emerging from these plants, but this apparently isn't done on a national basis. In consequence, water quality was estimated on the basis of such factors as the type of treatment plant (primary, secondary, or tertiary) and the number and type of purification pools. Because of these approximations, some cities may have either higher or lower indices than they deserve, but this was judged to be the only way of calculating these indices.

Perhaps the best way of making clear to the user of water indices the difference between those based on measurements and those derived from estimates is to segregate them by type. This was done in the Canadian index. Indices based on effluent estimates were considered in one group, and those based on actual measurements, such as mercury levels in fish, were put into another section. The overall water quality index was then calculated by using weighting factors to combine the two indices.

We shall never have perfect measurements. This means that we must continue to depend on at least some estimates of water quality. Since both of these approaches give us information on the water environment, we should use both in devising water indices.

THE RIVER BASIN—THE NATURAL GEOGRAPHICAL UNIT

One of the advantages water quality indices have over air indices is that there is a natural unit for the former type—the river basin. Without getting too technical, we can define the basin as the area from which a section of a river draws its water. To find it on a map, we draw the paths of the tributaries of the main river, then the tributaries of the tributaries, and finally enclose the veined area with a line.

Apart from the aesthetic pleasure of seeing the basins fitting together on a map, what advantages do we get from using them in water indices? Water quality of one part of a basin generally has a relationship to that of other parts. For example, when pollutants are dumped into a stream at one point, the water quality may still be poor miles farther down its length.

If no further harmful additions are made to the water, the natural "healing action" of the river will dilute the impurities, and the water quality index will fall to a more normal value. Examining the values of the index at points along a river in a particular basin shows that they are often similar. This is the main advantage of averaging water indices over a basin, rather than just showing them as unrelated points on a map.

Using basins to calculate indices isn't all sweetness and sparkling water. For example, we haven't said too much about the size of the river basins. The Mississippi–Missouri basin covers a large part of the continental United States. How is the water quality of a crystalline mountain stream in Colorado related to the muddy and polluted state of the Father of Waters as it rolls past New Orleans? Precious little—we have to draw our water basins much more compactly than this example. With sufficient care, we can subdivide basins into no more than a few hundred square miles. Such an area is likely to provide a reasonably homogeneous unit.

Another problem in using basins to determine average water quality is that of stream sequence. Suppose that streams A and B flow into C. A gets a heavy daily dose of municipal and industrial wastes, but B escapes unscathed. The water quality indices of the two will, of course, be considerably different. When the two join to form C, the water index of this stream will be in some sense an average of that of A and B. Although all of this is taking place in one water basin, we can't simply say that the index of A will be affected by what happens in B, or vice versa.

In spite of these and other drawbacks, averaging indices over basins incorporates many of the interrelationships among river flow rates, locations, and degrees of pollution. Why then don't we use analogous concepts for other environmental indices? One that has been suggested is the airshed in air quality. Air doesn't flow in channels as does water, but in some situations there are natural boundaries to air interactions. For example, in areas ringed by mountains, such as Los Angeles, the peaks form a barrier to modifications in air quality. We can think of the relatively flat part of the Los Angeles area as being a crude airshed. Enough mixing of air and its pollutants goes on to ensure the rough similarity of many air indices within this airshed.

Unfortunately, there are few such examples we can use. Air quality indices can easily improve (drop) by a factor of 5 in the space of a few miles traveled out of a city and into the country. If we knew more about air currents, we might be able to calculate the area of "mini-airsheds" over which the air indices are somewhat constant, but right now we don't have that knowledge. As a result, most maps of air quality indices show a series of dots, with no indication of the areas they represent, for individual cities and places.

WATER QUALITY INDICES IN FRANCE

The remainder of this chapter discusses water quality indices that are in use or have been proposed, with the advantages and drawbacks of each.

The French set of water quality data (published as *Inventaire du degré de pollution des eaux superficielles rivières et canaux,* La Documentation Française, Paris, 1973) is probably the most comprehensive yet published by a national authority. Strictly speaking, these are not indices, but values of water parameters and pollutants. All that is needed to turn them into indices are standards, which are available in the booklet that accompanies the detailed data. In effect, we have both the numerators and the denominators staring each other in the face. To calculate indices, we only have to divide the numerator by the denominator.

The French indices—we can rightly call them that—were presented in a series of about a dozen maps, each indicating one aspect of overall water quality. Only a few were based on estimates of pollution; most used actual measurements.

Let's take a few examples to see how the information was presented. One map shows the distribution of sulfates in water, measured at scores of locations. Values ranged from 0 to more than 1000 milligrams of the SO_4^{2-} ion per liter. This map gives us a good idea of the distribution of this pollutant. It should be noted, however, that some sulfates do occur naturally in water, especially in areas in which the soil contains gypsum.

In addition to the distribution in space, the way in which the values change in time is shown. This is one of the problems we wrestled with in discussing air quality indices. If we take a number of measurements of river water quality in the course of a year, should the value of the index be based on the average quantity, the highest (and most dangerous) level reached, or some combination of the two? The French index doesn't quite resolve this question, but shows both the high for the year and the average. Most places with high averages have high maxima, so the precise method of computation isn't critical.

A feature of the French index was the presentation of biological as well as physical and chemical indices. Other biological indices and indicators are discussed in later chapters. The index considered three microorganisms—*Escherichia coli,* fecal streptococci, and salmonellae—all associated with fecal contamination. The last two are more harmful bacteria and have a stronger resistance to water disinfection treatments. Shown are typical values of the counts of these organisms per volume of water, as well as the maximum values encountered over the year.

Another map shows the estimates of particulate matter put into the water by 27 major activities, ranging from the iron and steel industry to breweries. In addition, an indication is given of the amounts per inhabitant of a region. This last point may not seem significant, since the relative concentration of pollutant per volume of water, rather than per person, is really what we want. But if the pollution per head is high in an area, so may be the pollution per stomach. This area probably is the place to begin cleanup efforts.

The French indices go on to evaluate quantities like chlorides, BOD, and total hardness of the water. All of them suffer from one disadvantage: the indices are generally shown at separate points, rather than by river basin. The French efforts adequately show how water quality varies along short distances of a river. The price that must be paid for this precision is uncertainty about the state of regional water quality.

No attempt was made to use weighting to combine all the subindices into one national index, showing the overall state of water quality throughout the country. In other words, are the levels of *E. coli* more or less important than those of suspended solids? Until the relative significance of these and other water parameters is evaluated, both the public and policymakers will find it difficult to judge the quality of French rivers.

In spite of these problems, the French water quality indices probably represent the most substantial national effort to assess the state of this part of the environment. It will be a challenge for other countries to improve on them.

CANADIAN WATER QUALITY INDEX

The Canadian attempt at an environmental quality index contains a large section dealing with water quality. This work is divided into two parts. The first deals with estimates of effluents, or what was put into water. The second deals with what was actually measured in the water or in its fish.

The first subindex is in turn divided into five parts. The first is concerned with municipal sewage, and the other four with wastes from major industries, such as fish processing and pulp and paper. Since the industries discussed only account for a portion of the wastes going into Canadian rivers, this part of the index is incomplete.

An effort was made to combine mathematically the effluents from the various sources. For example, waste discharged into water from municipal sewage plants is considerably different from effluents of the petroleum refining industry. How are the differing relative toxicities of the pollutants accounted for? The Canadian water quality index used the standards set for each pollutant. For example, consider pollutants A and B. If A has a standard of 5 ppm and B has one of 10 ppm, it will take twice as much of B as A to produce the same toxic effect in the same volume of water. The higher the standard, the less toxic the pollutant. We can then enumerate a list of pollutants from industry, using the standards for each as a guide in computing an index of "equivalent effluents" for different sources. In this way the Canadian index, in a simplified manner, compares the relative effects of pollution from municipalities and four industries.

The subindex dealing with measurements of water quality is somewhat more restricted in scope, due to lack of measurements. The first of the three parts considers the levels of mercury in fish. Although this is also a biological index, it was placed with other water quality indices. The mercury index was calculated by dividing the concentration by the allowed standard, 0.5 ppm.

A second subindex evaluates trace metals in municipal water supplies. Over the past few years some scientists have reported unusual and potentially dangerous substances passing through our faucets. The data in the Canadian index are based on one of the few national surveys dealing with this question. Subindices that were prepared for metals like lithium and cadmium were combined to yield an overall index for trace metals.

The third component of the measured index concerns turbidity, or lack of clearness, of water. The more turbid the water, the less desirable it is for both human consumption and recreational purposes. The index of turbidity uses a dual scale designed to take account of both of these uses.

All eight parts were then combined by means of weights into an overall index. Its value affords a rough determination of the state of Canadian water.

There are some advantages to the Canadian index. The subindices are arranged in a hierarchical structure so that the importance of each to the whole can be evaluated. As well, many aspects of water quality are considered.

Perhaps the biggest drawback to this set of indices is that different scales were used for different subindices. Caused by inadequate knowledge of the effect of certain water pollutants, this defect should be overcome with improving knowledge.

Although not as comprehensive as the French set of indices, the Canadian work does take the somewhat daring step of trying to determine the water quality of an entire nation.

MANCHESTER WATER QUALITY INDICES

The Manchester environmental indices have been referred to in Chapter 3. Two of the six environmental indices deal with the state of rivers and canals. Canals crisscross the United Kingdom and Europe, although they are less common in North America.

The calculation of the indices for both the waterways was based on similar criteria. Four chemical and biological classifications of river water quality were combined into an overall index. The chemical classification is based on BOD and DO levels. The ability of the river to support different kinds of fish determines the biological classification. Only the chemical aspects were used to determine the final index, but there was strong correlation between them and those dealing with biology.

The rivers were divided into Classes 1 through 4. Class 1 rivers normally support game fish; Class 2 less prized fish, more tolerant of pollution; Class 3, even fewer fish; and Class 4 rivers do not support fish life.

To produce an average water quality index for each of the political subdivisions in the Manchester area, the river quality, as determined by this classification scheme, was calculated by the waterway length. For example, if there were 5 miles in Class 1, 3 miles in Class 2, and 7 miles in Class 4, for a total of 15 miles, the average classification would be $(5 \times 1 + 3 \times 2 + 7 \times 4)/15 = 2.6$. To convert this average into an index, the highest and lowest classifications of the 71 subdivisions were arbitrarily assigned index values of 0 and 100, and all other classifications were interpolated on this index scale.

At first glance it might appear that the Manchester indices are different from others mentioned in this chapter. To some extent this is true. The lack of adequate knowledge of how, when, and where to measure and interpret water quality leads to differences in the indices that reflect these measurements and interpretations. In spite of this, there are strong similarities among all water indices. For example, the use of BOD is common to the French and Manchester work. While effluent estimates were not used as a basis for the final Manchester indices, as they were in the Canadian work, they were discussed and considered. There are only so many types of judgments that can be made on water quality. Indices have to be based on some of them.

The question of effluent estimates versus actual measurements, which comes up so often in environmental indices, was dealt with in the Manchester study. In principle, if we knew the water quality of rivers flowing into an area, as well as the origin and quantity of effluents put into them, we would be able to calculate the average water quality in that region. Manchester work indicates that this can't be done with certainty. It was found, as might be expected, that the more effluents that municipalities and industries pour into streams, the higher the index for the water. The index can't, however, be predicted from the effluents. It's not sufficient to know that the factory a few miles away cheerfully flushes the equivalent of X units of BOD and Y pounds of acid down an outlet pipe each day. We need to measure their effect as well.

WATER INDICES IN JAPAN

We may have become too matter-of-fact about what water indices represent. Someone in a laboratory makes a physical or chemical measurement, compares it to a standard, and then derives the value of a water quality subindex. All done with clinical detachment.

Japan has had stark evidence of water deterioration. Perhaps the case that has received the most publicity has been the so-called *red tide*. The Seto Inland Sea, in the southern part of the country, receives a wide variety of pollutants from the heavy industry around it. While some organisms in the water are killed by the waste, others thrive. One type (marine euglene, a type of plankton) that has prospered is red in color. It has done so well that many parts of the Seto Sea have turned a bloodlike color. This in itself would be bad enough aesthetically, but the crimson organisms have also destroyed much of the remaining aquatic life in the area. The red tide is one of the most vivid effects of uncontrolled water pollution.

How do we go from scarlet shores to an index that can be analyzed dispassionately? One possible way would be to find the total shoreline of the Seto Sea (the standard) and determine how much of it is inundated by the red tide each year (the measured value). Dividing the latter quantity by the former yields a crude red tide index.

We can consider only a few of the many types of potential indices to which the Japanese studies give rise. They have given careful thought to their problems by setting up standards for a wide variety of pollutants. For example, effluent standards have been set up for substances generally recognized as toxic, including cadmium, cyanide compounds, lead, and hexavalent chrome compounds. These standards are generally around 1 milligram per liter. Other standards apply to BOD, suspended solids, phenols, copper, and a wide variety of other pollutants and parameters. With these standards, indices of water effluents can easily be devised.

In addition to effluent standards, Japan has devised ambient standards, dealing with what is actually in the water. The use of water must be considered when we set standards and calculate indices. For example, all water resources were divided into three sections—rivers, lakes, and coasts. These in turn were divided into the uses that were made of them. A total of 13 subdivisions were made and standards for pH (acidity), BOD, suspended solids, DO, and coliform bacteria set for each. By comparing the measured values to these standards, we can derive an index. For example, the average BOD level (in ppm) on the Tokachi River in 1970 (monitored at Moiwa) was 3.3. The standard was 3 ppm, so the index is then 3.3/3, or 1.1. The value of BOD on the Tosabori River was 33.3. Because this river's primary use was industrial, the standard was 10. As a result the index was 33.3/10, or 3.3. In a similar way indices can be calculated for sections of the same river that have different uses.

Sets of index tables can be used to develop maps that show how environmental conditions vary from place to place. For example, p. 127 of the 1973 Japanese annual report shows how chemical oxygen demand

(COD) levels vary in the Seto Inland Sea. The levels rise beyond 3 ppm. Since an appropriate standard for this type of coastal water is about 2 ppm, we can easily devise an index for COD. The index is greater than 1, indicating the standard is exceeded, near the industrial complexes of Osaka-Kobe (whose main production is machinery and steel) and Niihama (chemicals). We can then obtain possible correlations between the value of the index and the red tides.

Because of man's careful persistence in producing new products and carelessness in disposing of the wastes of his work, the list of standards must be continually updated. For example, the Japanese list of effluents has expanded from 3 in 1962 to 20 in 1971, and undoubtedly more are yet to come. Polychlorinated biphenyls (PCBs) are one of the prime candidates. This widely used chemical has turned up in the bodies of such disparate creatures as eagles and fish, and a worldwide program has been launched to discourage its production. Japan has set a tentative standard of 0.5 ppm for edible portions of fish and shellfish caught in coastal waters, and 3.0 ppm for those caught in inland waters.

Although the Japanese have made strong efforts to gather and interpret water quality data, their studies suffer from two problems. First, there has been no attempt to combine the indices to yield a fish-eye picture of water quality. Second, the river basin, the natural unit of rivers and streams, has been used only sparingly in showing the data. When these problems are overcome, Japan will be in a good position to act on the basis of the index results.

AMERICAN WATER QUALITY INDICES

Water quality indices in the United States are perhaps in a more formative stage than are those of the other countries mentioned in this chapter. One of the pioneering attempts to study the statistical aspects of water indices was by the National Sanitation Foundation (NSF), mentioned in the Thomas volume. We have been assuming, for the sake of simplicity, that a standard is a standard. However even experts can disagree. Part of the NSF study concerns the range of standards that would be chosen by people working with water quality. In addition, the experts were asked which specific subindices would be most illuminating in an overall water index. The exact statistical details are too complicated to go into here, but they clearly show that we still have some way

FIGURE 10
A water quality index must be based in part on chemical and physical measurements of water, but we ignore aesthetic factors at our—and the water's—peril. This oil slick near the Statue of Liberty is probably so thin that it would be difficult to measure using conventional techniques. However visitors to the symbol of America can see it all too well. Some water indices do include visual as well as physical factors. (EPA—DOCUMERICA, Chester Higgins, Jr., 1973. Courtesy U.S. Environmental Protection Agency.)

to go before we can set absolutely rigid measurement standards. As Figure 10 implies, setting aesthetic standards will be even more difficult.

Consider turbidity, or lack of clearness in water. This can be measured in Jackson units, which are a measure of the amount of light that can pass through a liquid. The NSF experts agree that if the turbidity is zero this aspect of water quality is perfect. In terms of the arithmetic used in this book, this corresponds to an index of 0. If the turbidity increases to a value of 30 units, the typical view of the experts is that the water quality has fallen to about halfway between perfection and worthlessness. However about one fifth of the group feel that at that value of turbidity the quality of the water is already worthless. Once we have an idea of the degree of difference among experts on setting water standards, we can work on narrowing the divergences.

The two main national water quality indices attempted in the United States have been limited in scope. The PDI index, developed by the MITRE Corporation, was based mostly on knowledgeable opinions of people responsible for river quality throughout the nation.

The first letter in the acronym stands for prevalence, or the number of miles of stream in the river being considered. The second represents the duration of the pollution effect. This point is usually somewhat neglected in the other indices discussed in this chapter. Pollution problems can occur either year-round or only during part of the year, because of changes in temperature or aquatic life. When a river is frozen over, we don't have to worry about its suitability for swimming. Most indices devised thus far don't consider the time factor adequately. The last letter in PDI stands for the intensity of the pollution. This consists of factors dealing with biological, aesthetic, and other components of pollution effects.

Since the PDI index involved opinions, not the comparison of measurements to standards, it is not directly analogous to the others we've mentioned. However, it did consider some factors that more advanced indices have neglected.

Enviro Control, Inc., under contract to the Council on Environment Quality, calculated what could be termed a protoindex for United States rivers and streams. Environmental indices have at least two major goals. The first is to determine the state of one part of the environment. The second is to tell whether it is getting better or worse. The Enviro Control effort focused solely on the latter goal, determining whether the measured values of seven groups of water pollutants and parameters, ranging from BOD to ammonia to suspended solids, had improved or worsened over a period of time. Of the seven, four remained about the same, one (suspended solids) improved, and two groups (total phosphorus, organic nitrogen, and ammonia; nitrogen trioxide and nitrogen dioxide) got worse. This approach still leaves us up in the air, or to put it more appropriately, submerged in the water, when we try to find the true water quality at the beginning or end of the time period.

Finally the massive MITRE compilation of 112 potential environmental indices, or "everything you wanted to know about indices," contains a sizable number dealing with water quality. These include red tides and algal blooms (mentioned in the Japanese studies), closed shellfish areas (included in the Canadian index but not discussed here), corrosion damage from water pollutants, depth of water table (included in the Japanese studies as a factor in land subsidence, but not discussed here), fish kills, and oil spills. The MITRE indices are often hypothetical, but they give an indication of the wide variety of possible indices.

CONCLUSIONS

The quality of the water that we drink and use for recreation can change our lives in many ways. Because of the wide variety of pollutants added to water, as well as the different manners in which they change water quality, work on water indices has not advanced as rapidly as that on air indices. Japan, Canada, the United Kingdom, and France have made starts on this work. We must have the knowledge provided by indices before we can gain a truer understanding of our aquatic resources.

6

LAND INDICES

To add just a bit to the song, "Fish gotta swim, birds gotta fly, man gotta walk. . . ." That walking is done, for all except Biblical prophets, on land. We've already discussed indices of air and water quality. Can we devise them for land, on which we spend almost all our days?

To compute land indices, we have to take a somewhat different approach from the one we had when we dealt with air and water. Most of our emphasis has been on the physical and chemical properties of these substances, and rightly so, because we can measure these attributes. We can measure some properties of land by instruments. We go into further detail when we discuss such aspects of land indices as quality of soils. But there are aspects of the land, for example, the extent and types of parks, that affect land quality yet can't be measured in the same way as the amount of DDT in the soil. As we develop land indices, we have to keep track of what can and what cannot be measured scientifically.

While an index of land quality is different from that of air and water, we haven't really decided whether it's truly necessary. I believe that it is. Anyone who has passed through hillsides denuded of trees, who has walked through the stark slashes of an eroded farm, who has contemplated scenic areas without adequate public access, knows that in some sense the quality of his environment has been lowered. It may not be easy or simple to devise a method of showing this, and the mathematics may be a little crude, but the attempt should be made.

WHILE STROLLING IN THE PARK ONE DAY. . . .

When we think of land of high environmental quality, we often visualize woodlands rolling up to mountain peaks, colorful deserts, or sparkling seashores. We want to be able to do more than think about them—we want to stroll through the forest, backpack along a trail, or splash in the surf. Since we usually can't do this if somebody else owns the land, what we're really thinking of are public parks.

City, county, state, provincial, national—the types and administrations of parklands seem to be beyond comprehension. When we add to the many types of park bureaucracies the vast variety of parks themselves, ranging from a few blades of grass and a battered picnic table, to some the size of some European countries, to yet others consisting of only sand and surf, we realize that the word *parkland* covers a lot of ground, both literally and figuratively. How can we possibly calculate an index with all these factors to consider?

The problem isn't insoluble if we make a few assumptions. First, some amount of parkland is better than no parkland. Second, a park that people can get to is better than one that is inaccessible.

Both of these propositions might appear to be self-evident, but the cold light of dawn might prompt some second thoughts. If our index rests on the assumption that having some parkland is better than no parkland, the index will also indicate that a big park is more desirable than a small park. This isn't always true. A small park may be expanded by adding land that is useless to everyone and everything. This doesn't necessarily make the park better. In general, though, the more area at our disposal, the greater the chances we have for our recreation and its conservation.

The second assumption may also lead to some nagging doubts. By now we've all seen pictures of the crowds at Yellowstone and other national parks, parked trailer hitch to trailer hitch and tent stake to tent stake. If this is what accessibility to parkland means, we may want to make ourselves inaccessible to it. The solution to such a problem is not to shut down Yellowstone, but to have enough parkland available near centers of population to banish the sardinelike scenes. Too many people tramping about can damage local ecology, but we should spread out the people, not banish them.

The environmental quality of land isn't the same everywhere, even in parks. Is there any way of quantifying this so that we can calculate indices? Perhaps the most successful attempt has been the work of Luna Leopold, who tried to rank the relative desirability of vistas and landscape types. He took into account such factors as the width of river valleys, the height of nearby hills, and scenic outlook. Chapter 8, which deals with aesthetic indices, has more to say about this. In spite of

Leopold's extensive labor, we still don't have a workable scheme for saying that an acre around Mount Whitney is worth more or less in terms of land quality than one in Central Park. Because of this, land indices assign both acres the same value.

Once we've made this reluctant decision, the next step is to see how we can compute an index for parkland. The obvious measure is the actual area of parks, whether city, regional, or national; what isn't so obvious is what to use as a standard. For example, in the Japanese studies the areas of national parks are carefully listed, but we have no indication whether or not the acreage is sufficient.

Part of this problem can be overcome by using the national average amount of parkland per person as a standard for use in calculating local and regional indices. For example, suppose that the national average of parkland is 0.1 acres per person. (We can, if we wish, express this standard as the fraction of all the land in the nation in parks, and arrive at similar conclusions.) If the corresponding figure for a particular region is 0.2 acres per person, this indicates that this type of land is relatively plentiful there, compared to the national average.

When we considered air and water quality indices, we usually divided the measured value of a quantity by the standard. If we follow this course for a parkland index, we get one running in the opposite direction from the others we've discussed. In the example we've used in the last paragraph, the index using this type of fraction is 0.2/0.1, or 2, which, according to our previous usage, indicates low environmental quality. The mythical area discussed really has high land quality in terms of parks, so we have to change the fraction around somehow. One way is to simply invert it, so that it becomes 0.1/0.2, or 0.5, which indicates a relatively good index. If another region has 0.4 acres of parkland per person, its index is then, under our new rule, 0.1/0.4, or 0.25, which is highly desirable.

A variation on this method of calculation was used in the Canadian work mentioned above. It found, as was expected, that the western part of the country had low indices of parkland, but parts of the more crowded eastern regions had comparatively high indices.

All of this may be very fine for those who live out where the skies are not cloudy (or polluted) all day, but what about the millions who live in cities? How do we devise an index of parks for them when even the most generous city provides only a small fraction of the acreage per person that a state or province can?

There are at least two ways to answer these questions. First, an index of city parkland has to take account of both the parks in a city and the accessibility of the nonurban parks to city dwellers. A city may have a fair proportion of its space devoted to parks but, if its inhabitants aren't reasonably close to national and regional parks outside the city limits, they are in some sense deprived. Putting this deprivation into the form of an index is a complicated mathematical task. We explore one method for doing this at the end of this chapter.

A second, more immediate response to these questions is to calculate an index of city parkland by a method differing from that used for regions such as states and provinces. In so doing, we're on safer ground, since many city planners have suggested standards for urban park areas. The details aren't discussed here, since planners often disagree among themselves on what the appropriate figure should be. Nonetheless, there has grown up in the last decade or so the feeling that every city should have a definite acreage of parkland per inhabitant. When some consensus is reached, we shall be able to avoid comparing cities to the national urban average, and employ a more systematic and thought-out standard.

Almost all of us believe that parkland is desirable. Differences of opinions enter when we try to decide how much is enough. One of the reasons we have to go through such mathematical calisthenics to get indices of parkland is this conflict of attitudes.

HAVING A WONDERFUL TIME—RECREATION INDICES

Parks vary in attractiveness to a camper, depending on whether they have 5 or 500 campsites. One park's uncharted bogs will likely lead a weekend stroller to favor another with well-marked trails. In short, there's more to a park than just its area. We need to devise indices that take account of the recreation possibilities of a park as well as its size.

These statements aren't completely contradictory to the reasoning of the previous section. True, we've described parkland and indices as if each park were the same except for area. This viewpoint was used for two reasons. First, it is simple. Second, there aren't mathematical ways of showing that park A may be more beautiful than park B.

Some halting steps have been taken to calculate indices based on the recreational potential of parks. They have been calculated for only a comparative few, due to drawbacks discussed later, but they're worth considering.

Perhaps the best analysis has been by Pikul, Bisselle, and Lilienthal, as described in the Thomas volume. Their work deals primarily with measures of recreation. Because standards are not usually included, we can't accurately call their results indices. However, if we wanted to produce indices, we could adopt the approach, mentioned in the previous section, of comparing a regional recreation measure to the national average.

Lack of places to walk in a park reduces its attractiveness for many people. We can then consider the total length of park trails and shoreline as one measure of recreation availability. When we divide the regional value of this quantity by the average shore and trail length per person in a nation, we have an index of recreation for different regions.

This measure still has its pitfalls. A trail that led through a field of thistles would attract only walkers with a leather wardrobe. A path curving around sheer precipices would have few takers, no matter how long or short it was. In other words, there are trails and there are trails. In spite of this, trail length is easily measurable and can provide a simple gauge of an important part of the land quality index.

Another obvious aspect of recreation capacity is the number of visits that are made to each park. Here the pitfalls start to gape wider. The number of visitors depends not only on how attractive a park is, but on other, nonenvironmental factors such as how much money they have to spend on travel, the availability of roads, and the time of year. For example, Canada has recently opened some magnificent national parks in the northern part of the country. Because there are no roads leading to them, the number of visitors is confined to those with enough money to charter an airplane—a small group at best. The scenery is reputedly breathtaking, but it won't show up in an index of park visits.

The number of visits to national parks like Yosemite fell strongly during the Depression of the 1930s. This was not because Halfdome had lost its charm, but because most people had less money to spend on vacations. The popularity of a park should tell us about the state of the land. It does, but sometimes the wrong things.

The report in the Thomas volume goes on to discuss other possible recreation indices, based on such factors as the diversity of facilities—can you ski, swim, and hike in the same park on the same day? are there mountain trails as well as those for Sunday strollers?—and the number of facilities—are enough campsites available? The ideas are still in the formative stage, and need clearer definition before they can be transformed into indices.

A recreation index is really just a park index that takes into account the different qualities of parks. We still have much work to do on both of these types of indices, however, before we reach the detail of air and water indices.

WHEN YOU CAN'T SEE THE WOODS FOR THE CUT-DOWN TREES—FORESTRY INDICES

The saying has it that only God can make a tree. Only humans, along with a few stray beavers and bolts of lightning, seem to have the inclination to knock it down. There wouldn't be much of a problem if reseeding, both natural and by man, proceeded fast enough to replace the trees mankind has burned or carted away. This has not occurred in enough regions, and by now most of us have seen areas shockingly denuded of trees and the wildlife they used to support. Describing an important part of the quality of land by a forestry index then seems natural.

Devising a forestry index isn't simple. From an airplane, the forest below may appear to be a uniform carpet of green. Approach it on foot and you find wide diversity, which should be reflected in indices. A sapling accidently chopped down is of much less significance than the demise of a towering redwood. Trees differ in species, size, age, and other properties. This leads to the conclusion that a forestry index can't simply be one number, but should consist of subindices dealing with different aspects of the forest.

If the woodsman hasn't spared that tree, forest environmental quality will be low. But if new trees spring up faster than the old ones are cut down, the forest will be in good shape. A possible measure of forest growth is the difference between the amounts chopped down and growing to maturity. If the area being cut (or harvested, using the euphemism of the paper companies) is less than the area replanted, either naturally or by man, forest quality improves; if it's the other way around, an index of forest quality gets worse, and the forest gradually disappears.

The Canadian index-makers ran into problems when they tried to construct an index based on these concepts. The difficulties were on two fronts. First, air and water indices generally consider only one quantity—the amount of pollutant. This forestry subindex attempts to evaluate the *difference* between two quantities, regrowth and cutting. Second, there is no easily definable standard for the amount of cutting that should be allowed. It would be all too easy to say, "Woodsman, spare all those trees." But this book is printed on paper, not air, and so is your newspaper. A thousand and one articles are made of wood and wood products, so the cutting is unlikely to cease.

Because of these mathematical quagmires, the Canadian index of forestry growth is not as effective as it could be. It shows areas where more trees are coming down than are going up, but the computational problems make the index difficult to combine with others.

A somewhat more successful attempt to devise a measure of forest quality could be called the Canadian Smokey the Bear subindex. Forest fires destroy the woods; if there are enough of them, one doesn't have to lift an axe or a chain saw to eliminate the trees. A simple measure of the amount of fire damage can be taken to be the number of acres destroyed by fire each year. A particularly bad one is shown in Figure 11.

Once again we're confronted with the problem of devising a standard. In this case, the concept of "maximum allowable burn," formulated in the scientific journal *Forestry Chronicle* in 1949, fits the bill. We can't give all the details involved in the calculation of this standard, but it takes into account such factors as the value of the forest for wood production, recreation, and wildlife; the ease of replanting after a fire; and the accessibility of the forest. The index is the actual area destroyed by fire divided by the maximum allowable burn for that type of forest.

It would be a temptation to stop here; after all, we have an easily measured quantity, the blackened area, as well as a standard. However, there's more to a forest fire index than the simple calculation we've sketched.

First, we should remember that, in spite of what Smokey has been telling us all these years, not all forest fires start with a match. Lightning started conflagrations before the slogan "close cover before striking" was ever thought of, and will continue to do so long after it's forgotten. In short, fires start by natural as well as man-made causes. Chapter 1 noted that environmental indices can be affected by both causes, and a fire index is a classic example of this.

Second, fires are *generally* bad for forests, but some are actually beneficial. (Will someone please hold Smokey still?) A well-known example

FIGURE 11
Indices of land quality take account of forests, one of our most valua-
ble resources. Part of the Canadian forestry index concerns the pro-
portion destroyed by fire, whether man-made or from natural causes
such as lightning. Although some fires are beneficial to certain types of
forests, most aren't. The Everglades will require years to recover from
the damage caused by this conflagration. To a forestry index, burning
can be just as destructive as cutting. (EPA—DOCUMERICA, Flip
Schulke, 1973. Courtesy U.S. Environmental Protection Agency.)

occurs in redwood forests. Redwoods are resistant to fire but grow
slowly. Unless a fire sweeps this type of forest every few decades, the
redwood saplings are crowded out by the nonredwood trees on the forest
floor, and the giants can't reproduce themselves.

As was mentioned at the beginning of this book, environmental indices are an approximation to the truth, not the whole truth. In the case
of the forest fire index, the approximations are bigger than usual.

A third forestry subindex that was attempted in the Canadian work
deals with the amount of forest destroyed by insects. If insects such as
the spruce budworm destroy the trees, the denuded effect that we see is
almost the same as if they were destroyed by man or fire. The quantity
that we then measure is the area ruined by insects. The lack of a standard makes calculation of an index difficult, but we can use the average
area destroyed by insects over a long period as a crude standard. When

these six-legged creatures go on a rampage, the index is high; when they lie low, the index does the same.

This would be a satisfactory index if it were never combined with any other. When it is, some inconsistencies come to light. One of the major ways in which insect depredations are controlled is by the use of pesticides. Chapter 5 mentioned that a possible water index is based on the concentration of pesticides. Later this chapter considers land indices dealing with pesticides in the earth. In other words, the chemicals so useful for lowering the index of insect damage may at the same time be raising other land and water indices. This makes at least two connected points. First, the environment is more than a word or a series of unconnected parts. All aspects fit together, and there are strong relationships between the parts. Second, we have to define environmental indices so that decreasing one doesn't inadvertently increase another.

The state of the forest seems so clear to us that we might think that devising indices to describe it would be simple. The Canadian effort on measures of forestry quality has been the most comprehensive effort yet, but our understanding of this field still leaves us at least partway up a tree.

STANDING ROOM ONLY—POPULATION DENSITIES

Excluding such small numbers of people as those who live in junks in Chinese harbors, most of us live on the land. And most of us live close to the rest of us. Surely a set of environmental indices must include one dealing with overcrowding.

Here we start to walk the thin line between social and environmental measures. It's true that we can blame overcrowding solely on mankind, whereas many environmental problems, as already noted, are only partly man-made. But almost all of the other environmental indices discussed in this book are based on physical, chemical, or biological measurements or calculations of one sort or another. None is based only on the presence of humans.

Nonetheless, when we're pressed to the back of a packed elevator or traveling through mile after mile of houses and factories and stores and office buildings, most of us feel that the quality of our environment is somehow diminished. How can we put these vague feelings of being crushed by our neighbors into an index?

We can measure the density of population easily enough. For example, the Council on Environmental Quality in the United States, in its fourth annual report, showed the population densities of three cities (Kansas City, Denver, and Riverside, California) as the distance from the central business district increased. The highest density of people is not in the middle of downtown, but from 1 to 3 miles from the center.

So we can measure population densities, at least on the average, by using census and other information. Note that we must emphasize the word *average*. We have no way of calculating the density of people at your neighbor's backyard barbecue or on the local freeway at 8:30 in the morning. Assuming that the population density is an average over these anomalous cases, we now need a standard with which to compare area to area and city to city.

Walk down the streets of a European or Asian town and you will note that most buildings are at least a few stories high and quite close together. The more outlying areas of the city usually show the same building patterns, in contrast to the suburban areas found at the same distance from the business district in most North American population centers. We haven't branched off into a short travelogue; we're merely emphasizing that what may be considered intolerable population densities on one side of an ocean may be regarded as perfectly natural on the other. What is acceptable in New York may be regarded as suffocation in Los Angeles. Any standard of population density we devise is probably not applicable on a worldwide basis, as are strictly physical or chemical standards.

In consequence, the Canadian environmental indices, the only set of indices yet that consider population, use as a standard the national average rather than a predetermined figure. The Canadian index uses crowding in persons per residential room rather than per acre because of measurement problems, but the two quantities are analogous. The index calculated showed that some cities, previously not suspected of having crowding problems, did indeed suffer from them.

For those of us who live in urban areas, the feeling of being continually surrounded by about half the human race is probably one of the ways in which our personal environmental quality suffers. Although it isn't simple to say just how much crowding is too much, any set of land indices should include an index reflecting population densities.

QUALITY OF SOILS AND EROSION

Although this chapter is devoted to indices of land quality, we seem to have talked about everything except the land itself. We've mentioned the trees growing on it, the parks we've fashioned on it, and the people crowding each other on it—but not a word about the very ground. Now is the time to rectify that oversight.

When we stoop down and let the sometimes-not-so-good earth trickle through our fingers, we may not have much of an idea of how to devise indices of soil quality. That trickling can give us a clue. Rainwater seeps through land. If the earth is eroded or not adequately held together by trees and plants, the water will wash away the land. It will eventually be deposited in rivers and streams, where we can measure the levels of sediment. Figure 12 shows that the earth isn't always as uniform as it looks on the surface.

The Canadian environmental indices contain one dealing with the sedimentation levels of rivers. In general, the more earth per unit volume of water, the more the best earth is being washed into the stream. There are at least three problems associated with this index.

First, it is difficult to say what proportion of the sediment level in a river is due to man-made causes and what, to nature. For example, when we cut down excessive numbers of trees and cultivate land that should remain prairie, as occurred in the Dust Bowl days, we cause the land to both blow away and wash away. But nature has a hand in the process too. Many of our rolling hills were once mountains, whose tops were flattened over the ages by the ravages of rain. The Canadian study showed that some rivers, though far from human habitation, have a relatively high index of sedimentation.

Second, some objection may be made to a sedimentation index because the measurements are taken in streams, not on their banks. It's true that the particles we find in the water came from the land, but the objection has some validity.

Third, we again suffer from the lack of an adequate standard. Our mistreatment of the earth we live on has caused much of it to be washed away, but we haven't yet determined how much we should allow.

The depositing of sediment in rivers occurs partly as a result of erosion of the land. It's then a secondary effect. Perhaps the logical step is to eliminate the middleman and look at erosion itself.

There is obviously a relationship between the two quantities. For

FIGURE 12
To a large extent, the value of an index of erosion is determined by the type of soil. Most of us don't have much knowledge of the earth beneath our feet until we see it cut open for a ditch or pipeline. (K. McVeigh. Courtesy Graphics Division, Environment Canada.)

example, the Soil Conservation Service in the United States has estimated that, of the 3.5 billion tons of soil lost each year through erosion from private lands, about 40% ends up as waterborne sediment.

The first of this pair of numbers illustrates, in an indirect way, the value of environmental indices. If the typesetter had slipped up and written 35 billion tons, and if the proofreaders hadn't caught the error, it's likely that nobody except soil experts and the Soil Conservation Service would have objected. If the information had been in the form of indices of erosion for areas across the country, it would have been considerably more understandable than these digits.

Comparatively little has been done to calculate erosion indices, even though the Dust Bowls of the 1930s, a product of erosion, were one of the most graphic illustrations of man's abuse of nature. The Canadian index attempted to calculate this index in an indirect way, since no measurements were available on the actual extent and location of eroded areas. Different soil types have different rates of erosion by wind and water, and these were tabulated. In addition, the amount of erosion is roughly proportional to the degree of use that is made of the land. In other words, it was assumed that an area with 80% of the land under cultivation suffers twice as much erosion as one with 40% if factors such as soil types are kept constant. The Canadian index obtained the areas of most probable erosion by multiplying together the factors dealing with degree of use and soil type. The western province of Alberta contains the areas having the highest values of the index.

This method of calculation is obviously crude. Ideally, we should know the true extent of the problem before calculating an index of erosion. Since we don't, we have to settle for the mathematically second best.

When we let the loam figuratively trickle through our fingers in the contemplation of a sedimentation or erosion index, we didn't pay too much attention to what was actually in the soil. Yet soil can be contaminated just as badly as water or air. This contamination can take the form of heavy metals, which are spread by industrial wastes through air and water, as well as agricultural compounds like DDT, aldrin, and all the rest of the chemical cupboard of pesticides and herbicides.

Japan has some of the most intensive agriculture in the world. This, coupled with the closeness of many industries to farming areas, has produced elaborate studies of soil pollution. As an example, Japanese authorities have monitored the concentration of metals such as copper, cadmium, and zinc in the soil. Many metals are found naturally in the soil, but the ordinary levels are much lower than when pollution is involved.

The Japanese offer us an example of how to construct an index of metals in the soil. They set a standard of 1 ppm for cadmium in rice. An area where the concentration in crops exceeds that level is designated a soil pollution control region. We can then compare the actual cadmium levels in rice (the measurement) to 1 ppm (the standard) to find an index. About 6% of the rice samples tested in Japan in 1972 exceeded the standard, or had an index greater than 1.

They devised an index for the cadmium levels in food rather than in soils because no standards have yet been set for the latter. When this is done, we shall be able to make better assessments of the quality of the soil.

The same reasoning applies to indices dealing with the levels of pesticides and insecticides in earth. We can measure pollutant levels in the soil, but this is difficult. Standards for acceptable human intakes of these chemicals have been suggested by experts appointed by the World Health Organization and the United Nations Food and Agricultural Organization. We can determine the average diet, work out the approximate levels of pesticides that we are consuming, and compare these to the international standards.

This was the procedure followed in the Canadian index dealing with pesticides such as DDT, lindane, and heptachlor. The actual measurements were made on food, but almost all of that food was grown on land. Let's see how the calculation was made. Every day we eat dozens of kinds of foods, all containing varying concentrations of pesticides. For simplicity, consider a strange and restricted diet, made up solely of cakes and ale. The rather tipsy person following this plan daily consumes 2 pounds of cakes, containing 5 ppm of DDT, and 3 pounds of ale, containing 3 ppm. The average level of DDT in the food he eats is then $(2 \times 5 + 3 \times 3)/(2 + 3) = 19/5$, or 3.8 ppm. The denominator in the fraction is the total weight of food consumed per day. If the standard for this pesticide were 2 ppm, the index would then be 3.8/2, or 1.9. We can then conclude that the acceptable level for DDT is being exceeded, as well as any possible acceptable level for carbohydrates and alcohol.

One of the advantages of measuring pesticides in food, as opposed to soils, can be seen by considering the ecological food cycle. Pesticides generally accumulate in higher concentrations as they pass up this chain. For example, the level of DDT in water may be of the order of a few parts per billion. Plankton and other small creatures pass the water through their cells. They are eaten by small fish, which in turn are eaten by larger fish. After the large fish are eaten by seabirds, the concentration of DDT in the bodies of the birds may be of the order of parts per million, or thousands of times greater than in the water. In an analogous way, plants growing in soil contaminated with pesticides can have a higher concentration than the soil itself.

In spite of our depending primarily on the land for most of our food, there are still many facets of it that we don't understand or don't measure. The indices we've discussed in this section, while admittedly imperfect, can produce somewhat better knowledge of the source of our daily bread.

DIG WE MUST?—INDICES OF STRIP MINING

In the early 1970s, we became aware of a crisis that seemed to match environmental problems in severity. At least one aspect of this energy crisis bears on land quality.

Although it's likely that we'll run out of oil and natural gas before too many more decades have passed, the reserves of coal, in North America at least, will apparently last for centuries, even at increased rates of usage. Those concerned with land quality find it annoying that coal often lies just below the surface. This location isn't annoying to coal miners, however, since they can easily scoop up the coal without having to build deep underground shafts. All very simple and economical—but in the process the land is stripped as bare as a banana before we bite into it. It is difficult to rank the various ways in which we ruin the quality of land, but strip mining must surely rank near the top.

In contrast to the difficulty of obtaining data on other aspects of land quality, we have heaps, if one can pardon the term, of information on the areas despoiled by strip mining. In the United States, for instance, the Earth Satellite Corporation analyzed the problem in a report to the Council on Environmental Quality. The subject was also included in the Canadian environmental quality indices.

Granted that it's difficult to hide a strip mine, what standard should we use in calculating an index? Those who would set the standard at 0 strip-mined acres would have some justification on their side.

Perhaps a more realistic standard would be related to the fraction of strip mines that are reclaimed to something approximating their original state. Manipulating the mathematics of this index to make it comparable in scale to other land indices might be difficult, but would be worth the effort. For lack of such a standard, the Canadian strip-mining index merely compared the average proportion of strip-mined area in different counties to the national average. This is a simple approach, but it does show the regions which have been chewed up the most.

There is little question that, as our supplies of other fuels dwindle in the future, strip mining of coal will increase in magnitude. If we can devise appropriate indices to describe it, we shall have a better chance of keeping its more destructive attributes in check.

DOWN IN THE DUMPS

An old cartoon shows a business executive staring moodily outside his office window. The sign on his office door says "Smith's Junkyard." His secretary whispers to a visitor, "He's trying to think of another word for it." Brood no more, Mr. Smith—now junk is called *solid wastes*.

Although some of our society's solid wastes are dumped at sea, most of it remains on land. How can we index the quantity produced? One way would be to measure its volume or weight for different municipalities and industries. This might be the most accurate method, but these measurements are very rarely taken because of the inconvenience.

Another approach to solid waste indices is mentioned in the Manchester study. The authors simply found the area in each of 71 political subdivisions that was covered by refuse, either municipal or industrial. The latter category is important in the United Kingdom because of the large amount of wastes from coal mining. In addition, the authors included land that was severely damaged by industrial development, called in the British terminology *derelict spoil heaps*.

An index was made from these acreages by calculating the fraction of land in each subdivision that was covered by dumps or otherwise made unfit. The region with the lowest value was given an index of 0, that with the highest a value of 100, and other regions the appropriate interpolations. In this way, we can obtain a rough measure of the fraction of land covered by, if not the actual amounts of, solid wastes.

There are a number of problems with an index of this type. For example, consider two municipalities having exactly the same volume of garbage. The one that spreads it a little more thinly will have garbage dumps that cover more ground than the other municipality's. Its index of solid wastes will therefore be higher. The method in which the index is calculated is an incentive to build high-rise dumps. To consider another example, suppose that one region contracts with a second to have most of its garbage delivered to the latter. The former area will then have a low index of solid wastes, no matter how much it produces.

Modifications of the index should be made when wastes are transported from one area to another, as they can be.

In spite of these and other potential difficulties, this index was the only one used in the Manchester study to describe land pollution, and is virtually alone in this field. While we may be able to determine how much of the land we're covering with our garbage, we have yet to decide what is an acceptable fraction. When we do, an index of solid wastes will become more meaningful.

'TIS A FAR, FAR BETTER THING—TAKING ACCOUNT OF DISTANCES

The influence of an environmental effect on us can depend on how far away it is. To a person on a photographic safari, a tiger 2 miles away is somewhat different from one 2 yards away. Can environmental indices take account of this type of distance effect? Yes, by considering both where environmental attractions (or problems) and the people are.

There are a number of ways of doing this. Before detailing the approach taken in the Canadian set of indices, we should note that considering distance effects adds yet another layer of mathematics to the computations. If the original data and assumptions to which this added stratum is added are sound, they will be able to stand the weight. If they are not, the entire structure is likely to come crashing down in a welter of numbers and algebra.

The mathematics used in the Canadian index, borrowed from the field of geography, is a part of what can be called *social physics*, in which some of the concepts of the physical sciences are applied to the social sciences. For example, suppose that we are in the middle of nowhere, a long distance from any parkland. We can then say that our accessibility to parks is very small. Another location could have Yellowstone just a mile or two away, Grand Teton a few more miles down the road, and a national forest not much farther than that. In this case, our accessibility to parkland is high. One way of demonstrating this mathematically is to assume that the accessibility falls off as the inverse of the distance from the park. For example, if we are 20 and 40 miles from parks of equal size, the relative accessibility to the first is twice that of the second. For the sake of simplicity, we're taking the distance as the crow flies, not as it walks, drives, or takes the train. When we plot all of the accessibility values on a map, taking account of all the parkland, we obtain a simple diagram of where the parks are with respect to each other.

But what about the people? We can pursue the same mathematical course, this time substituting the populations of cities for the areas of parkland. In addition, we should use the word *proximity* when applying it to people rather than *accessibility*, since the former term is more neutral. If you're in the middle of Times Square, you have more proximity to people than you do on a mountaintop in Montana.

Let's see how the relative values might be calculated. Consider the intersection of the 100th parallel of longitude and the 40th parallel of latitude, on the Kansas–Nebraska border. This point is about 300 miles from Denver, with a population of nearly 1.2 million, and about 200 miles from Wichita, with a population of around 400,000. If we assume that the proximity effect drops off inversely with distance, the average proximity is then

$$\frac{1,200,000}{300} + \frac{400,000}{200} = 4000 + 2000 = 6000.$$

When we add similar fractions for all of the major cities in the nation, we get a measure of how far (or close) this spot on the state border is from the population centers. We can do the same calculations for other places and then obtain a map of the population proximities.

The final step in the operation is to compare the park accessibility to the people proximity at each point. Someone near the Yellowstone gate has a high accessibility to parkland, which we can compute. He has low proximity to the people in cities, although on any particular summer weekend he might think that they've all decided to visit him. When we divide the proximity to people value by the corresponding value for accessibility to parkland, we obtain what we can call an index of accessibility to parks. A high value of the index for a given geographical point implies that parks are scarce and people are plentiful nearby; a low index implies the opposite arrangement.

This mathematical device enables us to obtain a clearer picture of where environmental amenities and problems are with respect to the population. It's not a substitute for more data or better understanding, but it puts what knowledge we do have in a clearer light.

CONCLUSIONS

The beginning of this chapter mentioned that, because we often don't know as much as we should about what we stand and live on, land quality indices are more difficult to devise than their counterparts for air and water. We have demonstrated that some indices can be calculated in spite of these problems. The scope of these indices has been broad, ranging from parkland to soil contamination. In addition, we have discussed a mathematical method for taking distances into account when we calculate these indices. Although it can't be claimed that indices of land quality are yet complete, definite progress has been made.

7

BIOLOGICAL INDICES

Hal, the talking computer in the science-fiction motion picture "2001—A Space Odyssey," charmed viewers. However, deep down he was only as good as his transistors and electronics. Although we're entranced by mechanical gadgets, most of us would probably prefer to watch the wigglings of an amoeba to the flickerings of a dial.

So it is with environmental indices. In one sense most of this book has been concerned with readings from scientific instruments and computers; important readings, but readings just the same. The effect of pollution on living creatures is what we're really interested in. A body of polluted water may send a needle flying across the dial of an instrument, but a far more graphic picture of the state of the water is the sight of dead fish floating on it. To show this phase of the environment clearly, biological indices are needed.

The two concepts are not mutually exclusive. We can have physical, chemical, and biological indices describing the same part of the environment. The type of index we choose will depend on such factors as the difficulty involved in making a measurement and the type of species affected. In some cases, a biological index will tell us more than any physical index can; in others, biological indices simply can't be devised. The remainder of this chapter is concerned, in part, with differentiating the two possibilities.

THE CONTRARINESS OF LIVING CREATURES—
MEASURING THINGS THAT RUN AWAY

As any fisherman knows, obtaining fish that can be used to devise a biological index is much more difficult than merely sampling the river or lake water. We encounter similar problems whenever we try to devise

FIGURE 13
These circling birds in Florida can disappear from sight with a flap of the wing. Indices of wildlife have to take into consideration that the counts of such creatures can vary drastically depending on location, time of day and year, and mood of the animals or birds. After what has happened to them over the past few hundred years, most species of wildlife are not eager to get close enough to humans to be counted. Counting problems must be beaten before we can devise accurate wildlife indices. (EPA—DOCUMERICA, Fred Ward, 1972. Courtesy U.S. Environmental Protection Agency.)

biological indices that deal with animals and other moving creatures. Plants, rooted to their clump of earth, are usually not too much trouble, but we're mainly concerned with the effect of pollutants on animals, fish, and birds. Most of these species are wary of us. So wary, in fact, that they make tracks, waves, or wingbeats whenever we approach. Figure 13 shows an example of this.

To a large extent, biological indices are based on the health and numbers of animals or plants. We can't base an index on the appearance of tail fins as the fish rapidly recede into the distance. Our relative unpopularity in the animal kingdom makes the compilation of biological indices that much more complicated.

Can anything be done to eliminate this problem? Wildlife biologists face this difficulty continually and have managed to cope. One of their duties is to estimate the population and vigor of the wildlife for which they're responsible—wildlife that often disappears at the sight of man. The task is challenging but not impossible. By using sampling and other statistical techniques, we can make reasonable estimates of the size and state of wildlife groups without counting every last hoof and horn.

So while biological indices are often not as accurate as physical or chemical indices, we can still get adequate measurements in some cases. For example, suppose that we're concerned with the health of fish in a lake. If half a dozen are netted at different locations and all have the same pollution-induced disease, we don't have to fiddle with slide rules to realize that we have a disturbing condition on our hands. Every fish in the lake need not be examined to come to this conclusion, any more than we would have to sample every cubic foot of air in a city to assess its degree of air pollution.

Some Biblical scenes in American Primitive paintings show the lion and the lamb lying down together beside Adam and Eve. Today, the lion, at least, would disappear at the sight of man. This doesn't mean that we can't devise biological indices based on the number of lions. What it does mean is that we have to make careful estimates before we can construct these indices.

BIOLOGICAL SENSITIVITY

We're interested in biological indices for more than the fun of counting animals or dissecting plants. These indices should give us information about the state of environmental quality not obtainable in other ways.

They do supply this knowledge because living things often are highly sensitive to pollution. This realization goes back hundreds of years, although present-day biological indices are relatively new. In the first underground coal mines, miners often perished from subterranean fumes. At the time, there was no way of measuring the levels of these underground gases, and the miners didn't care to act as biological indicators themselves. So down into the mines went a feathered pioneer of biological indices—a caged canary. This bird succumbs to fumes long before man does; that is, it's much more sensitive to pollutants than humans are. The sensitivity of another species is shown in Figure 14.

FIGURE 14
Fish are obvious candidates for biological indices. This scene in Chesapeake Bay probably shows one result of too much pollution. We can devise biological indices to warn us of what is happening before death occurs, so that we can take corrective action. (EPA—DOCU-MERICA, James H. Pickerell, 1973. Courtesy U.S. Environmental Protection Agency.)

Plants and animals may also have less sensitivity than humans, so we have to admit that we're leading with a trump canary. Some species of insects are so resistant to pollutants that they make humans look highly delicate. The progress of biochemistry has clearly shown which species have this tolerance. An article on biochemical indicators in the Thomas volume describes how some types of fish react to pesticides in water. The fathead minnow, for example, when exposed to the pesticide parathion at a concentration of 1 ppm for only 7 hours, loses about 15% of the activity of the enzyme acetyl cholinesterase in its brain. This biochemical change is easily detectable. Even more dramatic is the change in the bluegill species, which loses 85% of this enzyme's activity.

Contrast this biological indicator to an ordinary chemical measurement of water quality. We probably could measure 1 ppm of this pesticide in water without having to bring fish into the picture. However, it would be harder to show that one stretch of water had remained at that concentration level for only 7 hours—the time required to damage the brain of the fathead minnow. Because we can't measure all aspects of water in all streams at all times, our measurements are averages of sorts. A sampling of river water every day would be judged to be fairly frequent. In that time the 7-hour period of 1 ppm parathion could have come and gone without anyone noticing it except our submerged friends. Biological indicators can give us information on conditions that our instruments can't or won't detect.

An indicator isn't quite an index. To reach that elevated state, it has to be placed above a denominator—the standard. In the case of the brain enzyme, a maximum decrease of 10% has been suggested for acceptable fish health. Since the fathead minnow in the example we've cited had a 15% decrease, we can then say that the index is relatively high. In the case of the bluegill, with an 85% decrease, the index is extremely high. One way of calculating the index is to divide the actual decrease in the enzyme by the allowed standard. For the cases we've considered, the indices would be 15/10, or 1.5, and 85/10, or 8.5.

Our surroundings can be degraded by small, as well as by large, concentrations. For example, one of the most obvious pollutants is the amount of smoke and dust in the air, and efforts were started long ago to reduce their levels. But we now realize that minute concentrations of substances like DDT and other pesticides, entirely invisible, can do us perhaps as much harm as the all-too-visible grimy air. The instruments that we design, although sophisticated, often have difficulty in detecting these tiny concentrations. We can use the sensitivity of living creatures to indicate that some pollutants are around, even though at low levels. If we can also devise standards for these biological indicators, the resulting indices will be comparable to physical indices.

Biological and physical indices are complementary in many ways. We can use the former to measure low concentrations, and the latter to measure higher ones. By using both we can adequately cover all the stops on the elevator of environmental quality.

I'LL TAKE ONE OF THAT, THREE OF THOSE— BIOLOGICAL DIVERSITY

Suppose that we went for a stroll by a Kansas wheat field. We would see the stalks of wheat with heavy heads nodding in the wind, the sky, and little else. The number of other species, plant or animal, would be very small. To describe the scene in the jargon of the ecologists, we have an excellent example of monoculture.

Some scientists feel that monoculture, in even as specialized an area as a wheat field, is potentially dangerous because disease could wipe out crops that are all of one strain. If many varieties were planted, only one or a few at a time would be affected. The details of the logic are quite complex, and we can't decide here whether ecologists are right. Farmers might object, for instance, to planting corn every few rows in a wheat field. We'll let the farmers and the environmentalists battle over this one.

But there is a biological concept here that may be of help in setting up indices. That concept is *diversity*. Generally speaking, and we have to emphasize the word *generally* more than italics would do, the greater the number of species in a particular area, the healthier the area is from an ecological point of view. This bald statement makes all sorts of assumptions about natural changes in species due to climate and other factors. For example, the day before the swallows come back to Capistrano is not necessarily a day of low biological quality in that part of California. The birds migrate, as many others do over the world. If we were blindly following a rule of counting species to measure diversity, we would have to conclude that environmental quality improves suddenly on that day early in the year when the air is filled with beating wings. Migrant birds return regardless of man's activities. Biological and environmental indices are intended to give some kind of measure, however crude, of what we're doing to our surroundings. Of course, if we made life unpleasant for the swallows on Capistrano, by ruining their food supply, for instance, they might leave and never return. A measure of the types of birds that appear regularly in that area would then show a decrease, and we would be justified in saying that a biological index based on birds had worsened.

It hardly takes an ecologist to realize that natural communities of species interact in a complicated manner, even when man isn't around to throw a biological monkey wrench, in the form of pollution, into the works. For example, some bacteria reproduce every hour or so, and elephants every two or so years. This difference alone makes the ebb and flow of life in a meadow, stream, or savannah quite variable. The compilation of an accurate biological index based on diversity becomes more of a challenge. In spite of this, a diversity index can be valuable in telling us what we're doing to living things. If we increase air pollution, for example, certain highly sensitive lichens, scaly mosslike plants, may be killed or severely reduced in numbers. (These are discussed in greater detail later in this chapter). A diversity index based on the number of lichen species would get worse as the pollution increased, and we could use the values of the index as a guide to the effect of our activities. Similarly we can destroy some species of fish by putting chemicals and other wastes into the water, leaving only a few hardy species. The diversity index then shows decreasing environmental quality.

If diversity and other biological indices are so effective, why aren't they used more by governments? As was mentioned in Chapters 1 and 2, an official imprint on a scientific concept doesn't necessarily make it fit for human consumption, so to speak. Reports on hundreds of possible indices, both biological and nonbiological, have been published in scientific journals. An environmental index that is used by a government must have at least some degree of practicality. In effect, official agencies can perform a winnowing process, discarding the chaff of unsound scientific theories.

Governments have done comparatively little developmental work on biological indices as compared to the physical and chemical types discussed in other chapters. This is not to suggest that biological indices are inherently faulty or that there aren't enough biologists around to do the measurements, but rather that there is difficulty in interpreting the measurements. The problem of interpretation is associated with the setting of standards. Since nature is so variable even when man isn't around, a standard for the number of species in an index of diversity is not easy to set. An analogy might be a standard for the number of leaves on a tree. The number varies with the season, as well as with the strength of the wind. It might be possible to determine a standard value for each time of the year, but it would be highly changeable. The same situation applies to standards for diversity indices.

Some governments have made attempts to devise diversity indices. One of the more promising has been in Tokyo, for the river Tama. The number of species of water insects at a given point that could not withstand contamination was labeled A; the number that could withstand it was labeled B. The biotic index, as it was called, was defined as $2A + B$. The reasoning behind this choice of algebra was not made clear. Some drawbacks in the computation are immediately apparent. First, since there is no standard, we have some difficulty in relating the numbers we obtain to optimum conditions. Second, the mathematical scale of the index seems confusing. Suppose that we had 10 species of water insects existing naturally on the river. At a particular point, 4 are eliminated due to contamination. The value of A would then be 4; and the value of B would be 6 ($= 10 - 4$). The biotic index would then be 2 × 4 + 6, or 14. If more sewage were dumped into the river and only one species of insect could withstand it, A would be 9(10 − 1) and B would be 1. The biotic index would then be 2 × 9 + 1, or 19. The lowest possible value of the index would be 10, and the highest 20. These high values shouldn't cause us alarm. If a mathematically appropriate standard is used for comparison, the range of values can easily be made similar to that of other indices we've considered. We shouldn't be intimidated by the arithmetic if the indices show true environmental effects.

How can we make sure that indices of diversity are at least comparable to other types? The Tokyo diversity index showed one way of calculation. Perhaps a better method would be to use the number of species that were destroyed or significantly reduced as the measurement, and the original number of species as the standard. For example, if there were ordinarily 10 species in an area and human activities removed 2, the index would be $\frac{2}{10}$, or 0.2. Similar computations could be made for analogous situations. This calculation has the advantage that if no species were destroyed, the index would be 0, indicating perfection. The disadvantage is that its maximum value would be 1.0, less than the maxima of other indices.

The diversity of species in nature adds pleasure to a walk through the wild. It is possible to make measurements of this concept for use in an index, although natural changes do tend to obscure some of man's unhandiwork. An accurate index of diversity would undoubtedly be one of the more important biological indices.

WHERE THE DEER AND THE ANTELOPE DON'T ROAM—WILDLIFE INDICES

By now almost all of us have heard the term *endangered species*. As with so many other terms blithely tossed back and forth in newspaper headlines, many people don't understand the exact meaning. An endangered species can be described as one which has been so depleted in numbers that it has trouble reproducing enough offspring to avoid extinction. Perhaps the most prominent North American member of this unfortunate category of wildlife is the whooping crane, whose number in the mid-1970s was about 50. Even if we could guarantee that no whooping crane would ever again be killed by a hunter's bullet, it would be a long time before their numbers reached the few hundred necessary to avoid extinction by natural causes such as storms, disease, and lack of food. To put it another way, an endangered species is like a glass of milk on the edge of a table in a room with an active 3-year-old. Chances are that it won't fall off into extinction by itself, but it doesn't take too much of a push.

How are endangered species related to indices? The greater the number of endangered species, the more we have reduced the quality of our environment. We then have a measure that most of us can appreciate and identify with, in contrast to some physical measurements.

The term *endangered species* is simple enough to understand, but there are some problems with trying to devise indices based on it. First of all, how endangered is endangered? Some species, such as the whooping crane, are truly down to the last few feathers. Others are endangered only in certain areas, and abundant elsewhere. For example, the timber wolf and the buffalo have been exterminated in many of their natural habitats. They aren't doing too badly in others, with the overall result that they can't honestly be placed on a list of endangered species. Both aren't thriving as they were a century ago, but they're not on the brink of extinction.

To take account of this difference, the MITRE report on environmental indices suggested two types of wildlife indices. One would deal with endangered species as we understand the term, and a second with "troubled" species. The second category would encompass species that are depleted or rare. The exact meaning of the last pair of adjectives varies somewhat depending on the last ecologist you talked to, but they're similar to the description of the state of buffalos and wolves we've given above.

By adopting two potential wildlife indices, we can avoid some of the problems that are inherent in their use. A previous section in this chapter noted that in compiling biological indices we often deal with creatures that run away when we try to measure or count them. For example, the average timber wolf, the object of a bounty for years, is going to be highly vigilant of any two-legged animal. The wildlife specialist who goes out to count the wolves may not know one end of a rifle from the other, but the wolf isn't aware of this. The animal makes tracks; unless those tracks are in the snow, the counter often finds no trace and may conclude, unless he spots them from the air, that the wolves have been wiped out in his locality. Using two categories of wildlife indices allows us to put animals into a troubled species index until we're sure that they are really endangered everywhere.

The compilation of wildlife indices involves a philosophical point worth mentioning. Their calculation implies that we want to prevent every species from becoming endangered. This point of view is somewhat different from that which is often employed, where we assume that what affects mankind in the way of pollution or environmental amenities is the most significant point to consider. In calculating wildlife indices, we're saying, in effect, that what affects the so-called lower orders is significant. The two points of view don't necessarily clash—what harms man often harms animals—but in some cases there could be conflict. If we take the buffalo as an example again, probably more than $99\frac{44}{100}$ % of the numbers that roamed the North American continent two centuries ago have been eliminated by man. Most of this was done so that we could use their hides and populate the Great Plains without the shaggy animals chewing up the crops and grasslands. Man benefited, but the buffalo didn't. There's no objective way that we can resolve the conflict in cases like these. All that we can do is realize that it's a moral question. The sifting must be done, not through the earth we described in Chapter 6, but through our own consciences.

We've been implying that a wildlife index is confined solely to mammals and birds, by citing as examples wolves, buffalo, and cranes. Other living creatures can be endangered as well. For example, the Bureau of Sports Fisheries and Wildlife in the U.S. Department of the Interior classifies endangered species under seven categories: mammals, birds, fish, amphibians, reptiles, mollusks (like crabs), and insects. It's a little difficult for us to visualize an insect as being endangered, but the use of some of our more deadly pesticides can produce this result. All members of the animal kingdom, including man, can become endangered, and indices must reflect this fact.

Should indices of endangered species suppose that a whale and a mite have the same mathematical—not physical—weight? Some ecologists might favor this, saying that each species is of the same importance in the total web of life. Others disagree. For example, the MITRE study said that animals which are higher on the food chain should be given more weight in the calculations than those which are lower, since the changes in the former group can indicate effects that have occurred in the latter group. These changes might be, for example, an accumulation of toxic materials such as pesticides. The implication in the MITRE study is that the so-called higher orders are more sensitive to changes in their surroundings and food. This reasoning can be put into the form of rough mathematical weights. For example, the MITRE work on mammals gave a value of 5 to endangered carnivores, such as tigers; a value of 3 to ungulates, such as antelopes; a value of 1 to rodents, such as mice and rabbits. There can be arguments over the exact values chosen. In spite of this, the weights reflect, in a rough way, our attitudes towards wildlife. Most of us would be more concerned about the extinction of a species of leopard than a type of shrew. Although Dr. Doolittle might have found it easy to talk to the animals, he would have found it harder to rank them in terms of value.

There's a subtle point to avoid in considering indices of endangered species. An obvious measure for this index is simply the number of species falling into the endangered or troubled categories. Suppose that one of these unfortunates is the giant panda. Its numbers dwindle and dwindle until the last survivor disappears. The index then suddenly improves, since the number of endangered species has decreased. This clearly isn't what the index was designed for. To eliminate this anomaly, we have to add the number of species that have become extinct to those that are endangered. This was done in the MITRE study.

When we perform this operation, we are reminding ourselves mathematically that, although it's often all too easy to eliminate a species from the earth, all of the king's biologists and all of the king's wildlife experts can't bring it back again.

How can we calculate an index of endangered species? As we've mentioned, a measurement would be the number of endangered, troubled, or extinct species, not counting those that became extinct long ago. Setting a standard is another matter. The total number of species in the animal kingdom, including insects, is well over a million, and the number of endangered species looks very small in comparison to this. Perhaps a fairer standard would be the number of endangered species that were listed as such at some time in the recent past—say about two decades ago. Suppose that a nation had 10 endangered species in the mid-1950s. Since that time, 1 of them became extinct, 5 remained endangered, and 4 increased their numbers sufficiently to be removed from the list. An index would then be 6/10, or 0.6. In this example, the index could never fall below 1/10, or 0.1, taking account of the sole species that became extinct.

This type of index has obvious advantages, since in some sense it's an average of the things we are doing to the environment. In spite of this, governments have been reluctant to push forward on these calculations. For example, much information on endangered species comes not from official agencies but from conservation groups, such as the International Union for the Conservation of Nature and the World Wildlife Fund. One of the exceptions is the Bureau of Sports Fisheries and Wildlife in the United States, which has issued regular lists. Because of the complexity of determining which species are really endangered and which have become temporarily scarce, most governments have avoided measurements and standards that could be used to calculate indices. As a result we have little official data to present here.

Even the most polluted air can be restored to a pristine state if we turn off our dirt-creating machines. If a living creature is removed from the globe, there's no way that we can restore it. Indices of biological quality can take account of this fact.

LITTLE FISHES IN THE BROOK

Biological indices do not deal only with creatures on the land. Although lists of endangered species usually include aquatic animals, these are sometimes overlooked in our concern for land-dwelling ones. Since fish, mollusks, and other water animals and plants can serve as important indicators of water pollution, indices based on them are worth considering.

One simple aquatic index was used in the Canadian environmental quality index. When fish have more mercury in them than the allowed standard, the fishing season in contaminated areas can be shortened or canceled completely by government regulation. The shorelines where this had happened were shown in the Canadian work. This still isn't an index, however. Unless we know the total shore length, we can't tell whether 10, 100, or 1000 miles of restricted coast is a significant quantity. If the Canadian index had used as a standard the average length of shoreline from which fishing is conducted, the calculation would have more clearly shown the state of fish health.

An index based on the amount of shoreline closed to fisheries might more properly belong in Chapter 5, under water quality indices. Our object is not a rigid classification of indices, but a rough arrangement. Since the condition of fish was being measured, it seems more reasonable to place it in the category of biological indices.

The same logic applies to the Canadian index of actual mercury levels in fish, which is somewhat more direct than the shoreline index. This index indicates the levels of contamination in fish, rather than the geographical regions where they may be tainted. The standard was set by the Canadian government at 0.5 ppm of mercury, about half that set by some other countries. To calculate the index, the concentration levels in species like cod and haddock were averaged and divided by the standard. The places of fish catch weren't available, so the variation of the index from one part of the country to another could not be shown.

These two biological aquatic indices are among the few that have been calculated with official data. Many other possible types have been proposed by researchers in many fields and, while the methodology of calculating them hasn't been entirely agreed on, they're well worth considering.

If we want to determine the age of a person, we can look for visual clues. Older people are often somewhat stooped, move comparatively slowly, and have white hair—unless Miss Clairol has moved into the neighborhood. A lake has an age as well. It doesn't turn gray the way humans do, but it does get older.

Interested in the age of natural features like mountains or deserts? That's a job for geologists. Why should anyone else be concerned with the age of a lake? For one simple reason—if a lake gets old enough, it disappears. It doesn't vanish, leaving a giant hole where it used to be, but becomes covered by marshes and solid areas, which effectively end its use by many aquatic creatures. We measure the age of a lake by how much of it has disappeared.

Given enough time, many of our lakes will disappear completely. Others are being formed at the same time, so we don't face the prospect of having only dry land between oceans. But man often hurries the process of lake aging. Many parents attribute their gray hair to the riotous behavior of their children. We can trace the premature aging of lakes to man's often uninhibited actions. When we pollute rivers and streams, we may think that we're only discarding wastes. The pollutants are usually of little value to us, but chemicals containing nitrogen and phosphorus are often quite edible to microorganisms in the water. As a result of this continuous free lunch, algae and other plants undergo exuberant growth. Enough wastes can turn the entire lake into a mass of vegetation. This would occur naturally over tens of thousands of years, but our carelessness speeds up the action even faster than a Keystone Kop movie.

If we do nothing to stop this process, which is called eutrophication, what was once a lake eventually becomes essentially dry land. Since this is undesirable in most instances, the MITRE report proposes a set of eutrophication indices that would measure the health of lakes in terms of their biology. For example, the fraction of the lake that is covered by algae and other tiny but numerous organisms is a good indication of the amount of eutrophication. Other suggested indicators include the transparency of water. Usually the more living organisms in water, the less clear it is. Water turbidity, a related concept, has already been discussed in Chapter 5.

Most plants contain chlorophyll, which, among other properties, turns them varying shades of green. The amount of this substance— actually chlorophyll *a*—in surface waters can be measured. Again, the presence of too much chlorophyll indicates that the lake may be suffering from eutrophication.

The measurements of eutrophication that we've briefly described are only partially understood by scientists. As a result the setting of standards, which are needed before we can calculate an index, is in a rudimentary state. As we learn more about eutrophication, we should be able to construct useful biological indices describing it.

Some brides, if confronted by a thin bridegroom, will spend the first few weeks of their marriage stuffing him with food. If they are wise, they eventually realize that moderate eating habits are best, and taper off the whipped cream and potatoes. Lakes are often in a position similar to that of the bulging bridegrooms. Every lake must have enough food in the form of nutrients to feed its population of fish and other aquatic life. If we tip the balance by giving it too much, the lake can disappear, just as the bridegroom can disappear into a coffin through gross overeating. An index of eutrophication may eventually prove as handy as a set of scales for humans.

DEATH IN THE AFTERNOON—AND MORNING AND NIGHT

Even when eutrophication of a lake occurs quickly, it's still a slow process by human standards, taking place over years or even decades. Some events affecting the biological aspects of water can occur in hours or minutes.

One of the most prominent—and frightening—is the fish kill. If we pour large quantities of sewage, chemicals, or oil into the water, fish can be wiped out in massive numbers. In 1968 one overload of a petroleum company's waste treatment plant resulted in the death of an estimated 4,000,000 fish.

Two possible measures that we could use in indices are the number of incidents and the total number of fish killed. An equally obvious set of standards would be no incidents and no fish killed. Unless we consider events like the Biblical parting of the Red Sea, in which some fish may have been sliced in half, fish kills rarely occur naturally. However, once we start to calculate indices with standards set to 0, we quickly get into trouble. If we continue with the reasoning we've employed elsewhere in these pages, the standard is the denominator of a fraction. When the denominator is equal to 0, the value of the fraction is infinite, regardless of the numerator. While all of us, including the scaly victims, want to see fish kills eliminated from the environmental waterscape, it clearly isn't a simple matter to calculate an index based on these incidents.

LICHENS—LIKE 'EM OR NOT

By now you may have the impression that the only species worth worry-
ing about are members of the animal kingdom. Any plant would tell
you otherwise if it could. Plants, from the tiniest yeast to the tallest redwood, can be severely affected by pollution and so used in calculating biological indices.

There are perhaps tens of thousands of plant species, each of which
reacts in a different way to contamination. If one looks through the
scientific literature, one finds scores, if not hundreds, of types that have
been tested for their reaction to pollution. Many of these studies have
been fairly sketchy. From them we can form at least two conclusions.
First, biological indices based on plants are only in the formative stage.
Second, it may be useful for scientists to narrow their studies to a few
plants that are sensitive to air, water, and other types of pollution.

Let's look at a few possible plant indices. A tree in a forest may appear to be just that: a trunk, branches, and leaves. If you look a little
closer, you may find that parts of the bark are covered with tiny, often
colorful plants. The names of these obligate epiphytes are far larger
than the plants themselves, but we know many of them as mosses and
lichens. Since there are probably thousands of species of these plants, we
have to consider their overall properties, rather than those of particular
varieties.

Many epiphytes suffer from air pollution. Not in the same way as humans, though; none has ever been observed coughing or sneezing. The
symptoms they show are blight, disease, and death. If we can correlate
these symptoms with air pollution levels measured with the usual instruments, we'll have a comprehensive set of data on how this pollution
varies from place to place. You'll recall that we couldn't have an instrument on every street corner in a city, due to cost. By using epiphytes, we
are in effect putting an instrument on most forest tree trunks.

An epiphyte index has already been devised for Montreal. A
mathematical formula takes into account such factors as the number of
species, how much of the trees they cover, how often they are found,
their state of health, how easily they reproduce, and the strength of their
resistance to pollutants. Each epiphyte varies somewhat from the rest in
many of these factors, so a fair amount of arithmetical juggling has to be
done. The result of this work is an air pollution map of Montreal that is
more complete than one achieved solely by the use of scientific instruments.

This work hasn't produced an index as yet. To make one, we need a standard for the health of mosses, and this has proved difficult to formulate. Plants can look sickly even when they are many miles from the nearest source of air pollution. Houseplants suffering from human purple thumbs know the situation all too well. In addition, the absence of a particular moss from an area does not necessarily mean that air pollution has killed it off. There are no orange groves in the Arctic because of lack of sun and warmth; similarly, mosses may choose not to grow in certain areas because the available tree barks don't provide enough of the right nourishment. It takes experts to tell whether the absence of a moss is caused by air pollution or by tree incompatibility. Since there are only so many experts on epiphytes, it may be difficult to extend this Montreal protoindex to a large number of cities. Still, it shows promise for future understanding of air pollution effects.

Botanists have devoted much effort to identifying plant species that can be used to measure the effects of air and water pollution. For example, in the early days of the smog problem in Los Angeles, vegetation was almost the only indicator available. Plants have been used to show the effects of certain air pollutants, such as fluorine, that are only rarely measured by instruments. While we do not yet have a universally agreed-on index based on plants, the way has been cleared for its development over the next few years.

OTHER BIOLOGICAL INDICES

Because of the almost infinite diversity of plant and animal life, there are innumerable ways of devising biological indices. One of the more intriguing methods, which takes account of many rather than one or two species, is biomass indices.

Since this concept isn't very well known outside the fraternity of biologists, a few words of explanation are worthwhile. Suppose that you took one of God's little acres of land and weighed everything that lived either on or in it—the trees, the grass, the earthworms, the rabbits—everything. The total weight of all these plants and animals is called the biomass. We use this word instead of bioweight for linguistic accuracy, since physicists are only too glad to tell you that weight isn't the same as mass.

The biomass is a measure of the productivity of the acre. If man hasn't harmed the soil, polluted the air, or poisoned the water, the acre will probably have a relatively high biomass, provided that it isn't mostly desert sand dunes. Usually when we lower enviromental quality, we also lower the biomass. The more pollutants that we produce, the less chance living things have to grow.

Using this concept, we can produce a biomass index relatively simply, at least in theory. The measurement is the actual weight of all the living things in a given area. The standard is the weight of similar plants and animals in areas where the effect of man's activities has been negligible. We can't really weigh every last twig and beetle, so some estimates need to be made along the way. In spite of this barrier, a biomass index would be one of the few biological or physical indices that would integrate a wide variety of pollution effects. Because so few indices take account of more than one environmental factor, we should press forward on biomass and similar studies.

CONCLUSIONS

During the past few decades, a tremendous effort has been made to understand the life around us, ranging from work on the molecular structure of DNA to study of global distributions of plants and animals. Some of these labors have direct application to environmental indices. Although life tends to have much greater variability than nonliving systems, by using sufficient care we can construct biological indices on subjects ranging from endangered species to biomass. Since ultimately we are most concerned with how changes in environmental quality affect living things, biological indices offer great potential for achieving this understanding.

8

AESTHETIC INDICES

According to a popular melody of the 1960s, "The junkyard . . . and the billboards come between us." We can add to this song of unrequited love ". . . and the pizza stands, and the auto graveyards, and the parking lots." This may not add to the song's tunefulness, but it reflects some of the ways in which our environment is being degraded. These forms don't harm us chemically or biologically, as those described in previous chapters, but they do harm to our visual senses. This chapter explores some of the forms of aesthetic indices.

AT LAST, A USE FOR CRITICS

If you've ever been an actor or had a relative who performed on stage, you have some idea of the power of a critic. The review may be in an obscure school newspaper or thunder forth from the well-thumbed pages of a metropolitan daily, but it is read. When the evaluation of the performance is unfavorable, it is usually surprising how many people have noticed it.

Performers who have been verbally cooked often have the attitude, "What right does he have to criticize? If he knew what he was talking about, he'd be on the stage himself, rather than in front of a typewriter. For that matter, why have critics at all? Let people decide for themselves what they want."

Coming from someone who has been trampled by the critics, these words may be drenched in the juice of sour grapes. There is some validity, though, in the argument. We don't have to rely on critics to guide our judgments—partly because they can be so contradictory. Some claim that if you put three critics in a room, you'll come up with four opinions. Without some criticism of the arts, however, we would be at the mercy of the organizations with the largest advertising budgets. Although some criticism is ill-informed and biased, the level of taste of the population would surely sink without it.

The past few paragraphs may have read as if they were written for *Variety* rather than a book dealing with the environment. What is the connection between critics and indices?

There is little relationship between the indices previously discussed and critical judgment. No critic, however well-trained, can distinguish between 2 and 4 ppm of sulfur dioxide simply on the basis of his artistic knowledge. With aesthetic indices we may well need their understanding.

What we now have in the realm of aesthetic measures is opinion, not knowledge. Consider, for example, the Sunset Strip that lies on the outskirts of almost every North American town, with its neon signs, used car lots with flapping iridescent banners, and telephone poles winding their way through the maze. On one side of the issue we have the owners of the signs and hamburger stands, who, while they may wince a little as they go to work, defend to the last blink of neon their rights to catch the eye of every passing motorist. The other side consists of most of the rest of us, who wince more than a little. Some sort of critic, yet to be invented, who could make an aesthetic judgment, would be useful in these perpetual battles. Their chant could well be, "Lift that 20-foot-high bull off its lot / Tote that blinking sign away without fail / You get a streaming banner and you land in jail".

Almost everyone agrees that we have to have some forms of outdoor signs, if only to guide us to a public telephone so we can call in a complaint about too much advertising. The questions then are, how much advertising, and what kind. Some posters are not unpleasing to the eye. For example, a trip to Colonial Williamsburg, in Virginia, shows how advertising was handled in a relatively tasteful manner two centuries ago.

Although it's true that our cities and their outskirts can't all be modeled on the past, there are better and worse forms of commercialism. Aesthetic critics could give us a simple guide to which is which.

How do environmental indices enter into this controversy? Here we are entering—perhaps the words should be stumbling into—relatively virgin territory. These new critics should, with appropriate training, be able to devise scales that numerically rate different groups of outdoor displays.

A few hypothetical examples show how the calculation might be done. A street with unattractive storefronts, landscaping, and a Late Motley style might have a rating of 100 aesthetic units. This street acts as a standard of squalor. Suppose we consider a typical avenue or highway strip. That neon sign glares too much—add 3 units. That storefront is shabby and has cracked window glass—put on another 2 units. This would not necessarily be the actual point system, but any aesthetic index would have to use similar procedures.

Is the scaling similar to that of the other indices we've considered? After the first neon sign the aesthetic value is 3; after the shabby storefront the value is 5(3 + 2), and so forth. The more affronts to our vision, the higher the index. When the eyesores are eliminated, the index is 0. If we divide by the standard of 100 units, the scale can become comparable to other environmental indices.

We're clearly tiptoeing through the artificial tulips here. There are no aesthetic critics authorized to cast a jaundiced eye, and there is no clear way of training them. Yet most of us have some idea of what constitutes a visual atrocity, and most of us want to get rid of them. Environmental indices may be one way of ranking man-made cityscapes. Since we don't have complicated instruments for measuring the state of the visual environment, as we do for air and water quality, we may have to train aesthetic critics to do this work.

HIS YARD IS MESSY, MINE IS ONLY LIVED IN

It all sounds admirable—squads of critics roaming the land, viewing with dismay and pointing with pride, then incorporating their findings into indices. However, there are likely to be more problems associated with aesthetic indices than with any others in this book.

Difference of opinion is one of the main reasons that it's sometimes difficult to get along with fellow humans. The matter can be illustrated by the murmured slogan wafted over suburban fences in summer: "His yard is messy, mine is only lived in." When we translate this attitude to aesthetic indices of commercial areas, it becomes, "His orange sign is a disgrace to the neighborhood; mine, in a delightful blue, is a credit." Figure 15 shows the differences in perception that can arise.

Is it possible to agree on aesthetic indices in spite of our fondness for disagreement? The problem isn't as insuperable as it looks. There is

FIGURE 15
A boring stake fence, or a mysterious and beautiful sunset in California? Differences of opinion complicate the job of devising aesthetic indices. However, most of us do agree on what constitutes an eyesore, and we can use this agreement to devise indices of what we see, as well as those of what we breathe and touch. (EPA—DOCU-MERICA, Tomas Sennett, 1973. Courtesy U.S. Environmental Protection Agency.)

often more agreement among critics than you might think. Let's illustrate it by an example. Suppose that your Aunt Minnie is a weekend painter. If she's like most others of this type, she tries hard but doesn't quite succeed artistically. You take her oil paintings to a set of critics, ranging from a neighborhood art dealer to university faculties. Unless you've got another Grandma Moses on your hands, most of the critics will be in agreement that Aunt Minnie should stick to her easel for the fun of it, not for the artistic success. This seems to indicate that aesthetic indices have some chance of seeing the light of day.

Even among scientists there is more disagreement than meets the layman's eye. True enough, most of the differences do not concern the fundamentals of science, but the smaller points. Regardless of the size of the differences, there are enough varying opinions of what the facts are to provide a few raised voices at scientific conventions. In particular, some of the environmental points raised in previous chapters are by no means as settled as an old dog by a fireplace. Environmental standards can and do change as more scientific facts come in; there's no reason why aesthetic standards can't change as well.

All is not sweetness and light in the land of the evaluator. Most of us, at one time or another, have read of the titanic struggles in newspapers and art journals between groups of critics about this or that new sculptor or painter. Does this mean that the critics will work themselves into a pitched battle over Aunt Minnie's brushstrokes? Not likely. Most of the controversy arises over how new work should be evaluated. When they first appeared on the scene, the French Impressionists were the subject of violent critical clashes, although now they're regarded as old hat or, in the vernacular, vieux chapeau. The same process has been repeated for action painters such as Pollock and pop artists like Warhol; arguments will someday undoubtedly rage over the quality of pictures that are painted by eyelashes. But Michelangelo and da Vinci don't go in or out of fashion. However many the waves on the surface of the ocean, deep down the water is fairly calm. If somebody invents a new type of neon lamp that is less bright than the present models, our corps of aesthetic critics may disagree on its value to the visual cityscape. However, they're likely to be in agreement on most of the familiar parts of the town scene.

Indices of urban aesthetics will be difficult to construct. Now there are only vague ideas floating through the air, not specific blueprints. Because there is a wide area of agreement on the quality of what we see, setting standards won't be impossible. If we are going to improve our aesthetic as well as our physical environment, we must develop these indices.

HOW SWEET IT ISN'T—INDICES OF ODOR

Faster than a speeding bullet—that's the way the coming of Superman used to be advertised. More sensitive than a complicated air quality instrument—that's the way our noses could be touted.

A logical place for a discussion of odor indices might be in Chapter 2, along with other air indices. However, what we smell is often not quite the same as what our measuring instruments detect. For example, although carbon monoxide and lead are important parts of an overall air quality index, we can't smell them. On the other nostril, we can smell some of the hydrocarbons emitted by auto engines.

Indices of odor are discussed in this chapter because they're more than just air indices—they're connected with the aesthetics of air. Nobody is going to rank air on the basis of how it looks, unless they're in downtown Los Angeles, Mexico City or New York. But its smell is a part of its quality.

If the ordinary—perhaps not so ordinary after all—nose can detect pollutants other than those discussed in Chapter 2, is our understanding of air quality somehow defective? Not really. We can think of the nose as an instrument able to detect some pollutants that physical instruments can't. To a nose these pollutants are literally a breeze.

Constructing any aesthetic index is a challenge. Odor is no exception. Although most people can describe smells such as musty, putrid or burnt, putting this knowledge on a scale is not simple.

The nose is sensitive enough but, like all human senses, it can be fooled. In an extensive review article in the Thomas volume, Engen recounts the story of some people who lived beside a Swedish pulp mill, known for its strong and unpleasant smells. The surrounding population used to smell the chemicals continually—even when there was no odor coming from the mill. It used to be said that the heart has a mind of its own. We now know that the nose has a memory of its own.

We've been talking as if the human race had only one collective nose. There are billions of noses, each with its own different sensitivity. You may gag at a very low concentration of menthol smell. I may be able to walk obliviously through clouds of menthol but be highly sensitive to a lemon odor. One might think that the way out of this confusion and into the realm of agreed-on standards might be to have a panel of trained human noses. This panel could consist of people, such as perfume testers, who earn a living by their sense of smell. However, Engen indicates that these professionals do little better than untrained college students when it comes to detecting and estimating the strength of industrial smells. The perfumers may be good at detecting the difference between Chanel No. 5 and Guerlain, but when it comes to the pollutant smells, we all must sniff for ourselves.

Because we are so variable in our ability to detect and estimate the odor concentrations, there has been great difficulty in devising indices based on them. The problems lie both in setting standards and in devising a reasonably accurate system of measurement.

In spite of these intricacies, Japan has made some progress. Because of the care that went into this work, it's worth discussing in some detail.

The Japanese work was divided into two parts, dealing with effluent estimates and measurements. As you'll recall from other chapters, most environmental information can be separated into these two divisions. The more smells we put into the air, the more our noses will wrinkle up, although the relationship isn't exact.

Let's first consider possible indices of odor effluents. The Japanese identified five substances that were to be controlled: ammonia, methyl mercaptan, hydrogen sulfide, dimethyl sulfide, and trimethylamine. It would be difficult to describe each of these odors without attaching to this page the little rough squares sometimes seen on magazine advertisments for perfumes. When the square is scratched, some of the perfume being described inadequately in print is released. If this technique had been adopted here, this book would be banned in many more places than Boston.

Words alone are inadequate for conveying impressions of smell. Of the five odors described, ammonia is familiar to anyone who has ever mopped a kitchen floor. Hydrogen sulfide has the approximate smell of rotten eggs. Some butyl mercaptan can be found in an area where an enraged skunk has stormed by.

Foul odors can come from anywhere, but the main source of the five odors mentioned above is industry, in the form of plants processing everything from feathers to artificial fertilizer. The Japanese have set standards for each of the odor pollutants, based on estimates of how much wafts over the factory fence. In turn, these estimates are based on the height of the stack through which the pollutant passes, and a factor based on the relative potency of each odor to human nostrils. Because we often don't know how odors are generated, dispersed and carried to our noses, these standards are crude. For example, not all odors from a factory conveniently pass through its smokestacks.

There are many more than five industrial odors. Some experts have stated that there are hundreds of distinct smells that can be offensive to humans in varying degrees. Since the chemistry of most of these stenches is often complicated, the Japanese work concentrated on some of the more prominent and easily identifiable ones.

How do we devise an odor index from these standards? For a particular factory, we can compare the actual flow rate of the smell in question to the standard that was set on the basis of the stack height and other factors. For example, the ammonia standard for a factory with a stack height of 20 meters (about 60 feet) is the discharge of approximately 40 ppm of this pollutant as a proportion of the total gases produced. If the actual concentration were 80 ppm of ammonia, the index would then be 80/40, or 2.0. The environmental quality of the air around this factory would tend to be poor if the prevailing winds were distributing the smell fairly uniformly.

Considering the effluents of odor from a factory is only one way of devising an index of odor. We eventually must measure to determine its effects on human feelings and health. For want of a better instrument, our noses can be used. Most human senses are difficult to quantify. For example, we can usually tell that one light is brighter than another, but we often have trouble saying by how much. In spite of these problems, which are even more pronounced when it comes to our sense of smell, the Japanese constructed a scale of odor intensities.

The scale was divided into six parts, ranging from 0 (odorless) to II (slight odor) to VI (extremely strong odor). Exactly how the judgments were to be made was not clarified. Even if factories were belching out noxious fumes day and night, a rash of the flu or head colds might keep nearby residents relatively content. They could also get used to the stench. However, the scales do provide some means of judging odors.

Let's see how we can devise odor indices. Suppose that we took level I—barely detectable—as a standard for a particular odor such as ammonia. People have different sensitivities to odor, and we can use this fact to construct an index. Suppose that we had a panel of 10 noses, in which 5 detected no odor (level 0), 3 barely detected it (level I), 1 detected a slight odor (level II), and 1 highly sensitive person detected a strong odor (level IV). If we substitute Arabic for Roman numerals and multiply the number of noses by the level they detect, the average level is then $(5 \times 0 + 3 \times 1 + 1 \times 2 + 1 \times 4)/10 = 9/10$ or 0.9. The denominator in the fraction is the number of people (or noses) in the test panel. The value of 0.9 is then a measurement, and its units are in levels. To produce an odor index, we divide by the standard, level I, which by our transmutation of numerals corresponds to 1.0. The index is then 0.9/1.0, or 0.9, which corresponds to a relatively high value. If other smells were present, we would undoubtedly obtain different values of the index.

The nose knows how to detect smells, since that's one of its main functions. Putting this sense on an index basis can be a formidable task, and efforts to do this have only started. Since our aesthetic environment consists of what we smell as well as what we see, an index of odors must be part of the evaluation of our surroundings. To smell or not to smell—that will continue to be the question.

AN INDEX WE CAN'T REFUSE—GARBAGE AND OTHER LITTER

Most people have probably come across more than one scene in nature that was close to being perfect—except for the evidence that man had been there. Squashed beer cans, bits of facial tissue, smashed bottles, and scraps of newspapers clutter all too many landscapes. This junk is clearly part of our aesthetic environment. Is it possible to measure the mess we've made of our surroundings?

There's no simple answer to this question. The old anti-dumping slogan, "Every litter bit hurts." assumes that the more waste material on the ground, the worse the situation is. Then every piece of scrap paper counts equally, whether it's white or a garish shade of orange. This may be a rather crude system, but we have no objective way of determining whether a rusted can is uglier than a tattered shoe.

Chapter 6 discussed garbage from a slightly different point of view: the amount of land covered by refuse from municipal and industrial sources. The present section considers the aesthetic quality of littered land, not including dumps. Most of our garbage goes directly to these dumps. The relatively small amount that we scatter about can be much more offensive to our senses than the rest.

A measure of litter is no simple matter. One of the few that have been proposed is that in the MITRE report, where the number of bags filled with garbage per mile of highway is proposed as a measurement. The most obvious drawback is that by using this reasoning we would be measuring volume, not unsightliness. Figure 16 shows some of the volumes a big city generates. For example, a discarded mattress would get more emphasis than a hillside of pink Kleenex. The road method also doesn't take account of parks and playgrounds, whose litter may have to be calculated per acre rather than per mile. Still, for lack of any other measurements, the sheer volume of our wastes may prove adequate as a first approximation to the truth.

A standard for litter could be none at all, but it's unlikely that the

FIGURE 16
An index of land covered with garbage and refuse could be improved by towing the wastes out to sea, as shown in New York City. But any index of ocean wastes would show deterioration under this scheme. We can't eliminate an environmental pollution problem by moving the wastes elsewhere. Bathers near New York who swim in floating garbage may not feel that the index of refuse had been improved too much by these towboats. (EPA—DOCUMERICA, Gary E. Miller, 1973. Courtesy U.S. Environmental Protection Agency.)

130

human race will collectively be as neat as third-grade teachers. There will always be a few gum wrappers blown out of the hands of little children. We may have to leave the setting of litter standards, without which litter indices would be difficult to compute, in the hands and minds of the aesthetic critics we referred to earlier in this chapter.

Man has often been described as the messiest of species. It's difficult to know whether this is true, although one rarely sees dogs or cats throwing orange rinds back over their shoulders. An index of litter, while not resolving this question, can at least show which parts of the nation have suffered the most from our barrage of waste.

NATURAL AND MAN-MADE BEAUTY

To judge sights on the basis of pleasantness to the eye, we have to have some idea of what is attractive. Philosophy professors may trot out hoary maxims such as "beauty is in the eye of the beholder" and "beauty is truth," but this won't develop an index. We need a method to measure the ways in which we see nature. Only then can we feel free to set loose our battalion of aesthetic critics.

Although bookshelves overflow with tracts on what constitutes an attractive landscape, comparatively little has been devoted to analyzing it critically. Perhaps the most detailed study has been one by Luna Leopold on 12 river sites in Idaho, published as *Quantitative Comparison of Some Aesthetic Factors among Rivers,* Geological Survey Circular 620, Washington, 1969. Because of its complexity we can only give a bare outline of its reasoning. The study was confined to river landscapes but could be extended to others more familiar to us. A total of 46 factors affecting the landscape were considered. These were grouped into three sections—physical factors, biological and water quality, and human use and interest. Each of the 46 factors had five possible categories into which the landscape was fitted.

Some of the factors come from the sciences of hydrology and geology, such as the river width, depth, and velocity. A few, such as water turbidity or lack of clearness, are discussed in Chapter 5, which deals with water indices. Still others, such as the number of algae in the water and types of land plants, belong in the realm of biological indices and are evaluated in Chapter 7. But Leopold does consider many factors that are solely aesthetic in nature.

Let's consider a few. Some, such as the number of occurrences of trash and litter, were considered in the previous section. Others, such as the degree of accessibility, are more general measures of aesthetic quality. This measure ranged from wilderness, which received the highest rating, to urban or paved access, which received the lowest.

Most of us have oohed and aahed at views from hilltops and mountain peaks. Leopold included this factor, which ranged from vistas of far places (the best), to no vistas (the worst). Many sites having high ratings in terms of other aesthetic qualities ranked low on this scale because of the way the hills and trees obstructed the scenery.

Many other factors were mentioned, including the potential for recovery from man-made changes, historical features, obstruction of views by electric utility lines, and the degree of urbanization. Although it may never be possible to capture all of the aspects of landscape beauty in a cold numerical table, we can evaluate many of them.

An area with beautiful scenery pleases our delighted eyes, but how do we define it in terms of numbers? Leopold attempts to do this for certain factors, such as the scale of the landscape, by means of graphs. These are too complicated to discuss here, but take into account the height of nearby hills and the width of the river valley. Only a few factors are treated in this mathematical way, but it does show a way around the problems of subjectivity. As long as you, I, and the person down the block have our own personal opinions on how our surroundings look, we're going to differ on what we see as beautiful.

The Grand Canyon in Arizona and the Grand Teton Range in Wyoming have more than beauty to recommend them. They are unique on the continent, and some of their qualities may be unique in the world. Is it possible to evaluate novelty in a list of aesthetic factors? Leopold does this by considering how many of his group of 12 rivers fall into a particular category for each factor. For example, if only 1 out of the 12 had a value of 1 for trash and litter (indicating little or none), it had a high uniqueness ranking. Unfortunately, this mathematical procedure was followed for other values, so that, if a river had an amount of trash far exceeding the others, it was also deemed unique. The Little Salmon River in Idaho, which is a sluggish, algae-infested, murky stream, had the greatest degree of uniqueness—only in the wrong direction. The calculation of novelty must be improved before it is satisfactory.

Implicit in the discussion of these schemes for calculating aesthetic quality is the assumption that beauty can be divided into factors. Frankly, there is no way that we can prove this claim. We can't measure the beauty of the Venus de Milo by counting the number of arms, but we can try to evaluate natural landscape by counting the scenic vistas and the beer cans.

Mathematical aesthetics is of interest, but is it an index? Since each factor for each river has a number attached to it, we can easily calculate an index. To do this, we also need to assign weights, so that the relative importance of the different factors is considered. For example, is the number of utility power lines across the land more important than the accessibility to it? We still don't have answers for questions such as these. If we assume, for the sake of simplicity, that the factors get equal weight, we can construct a landscape index. The Salmon River near Riggins, Idaho, had a value of 1 for trash and litter, 1 for accessibility, and 2 for utility power lines. The average value is $(1 + 1 + 2)/3$, or 1.33. With what can we compare this abstract number? We suffer here from a lack of standard. However, the relative value of this protoindex is still lower (better) than that of the Warm Lake site on the Salmon River, which is about 2.8. Two objectives of environmental indices are to compare one area to another and to compare each area to a comprehensive standard. We may have to be satisfied with achieving only the first of these objectives.

Although this evaluation of landscapes is not by any means complete, it illuminates the path toward greater understanding. Indices that describe the state of our natural landscapes should enable us to better measure man's effect on his surroundings.

CONCLUSIONS

Aesthetic indices are perhaps the most difficult of all to visualize and formulate, if only for the reason that each of us has a different view on what constitutes beauty. The problem is compounded because we attach varying importance to each aesthetic aspect. We have demonstrated that in spite of these problems we can devise indices describing such diverse qualities as litter, odor, and landscapes. For many people, indices of aesthetic quality are the most important of all, since they deal with things that they can see, touch, or smell. Three out of five senses is a significant proportion. When environmental scientists realize a little more clearly the significance of aesthetic indices to the public, we may be able to make more progress on them than we have to date.

9

OTHER ENVIRONMENTAL INDICES

Every child, as well as most adults, has faced the problem of trying to put square pegs into round holes. During my short sojourn in the armed forces, our compulsory chapel services were divided into the main religions: Catholic, Methodist, Baptist, Jewish, and so on. This left a rather disparate group of atheists, agnostics, Zen Buddhists, and splinter sect fundamentalists loosely grouped into "Others." We shall have to adopt the same convention for environmental indices. Some of the indices considered in the next few pages don't fit easily into the other chapters, although they deal with important aspects of our environment. A few are partly environmental and partly social. All of these can be grouped together under the gloriously broad term of "Others."

RAYS FROM ALPHA TO GAMMA—RADIOACTIVITY INDICES

If you've ever stumbled around blindly in a darkened room, you know that the objects you can't see are more frightening than the ones you can. A chair that seemed innocent enough during the day seems intent on kicking your shins in the dark. All of the furniture has become diabolical because one of our more important senses, vision, is lacking.

The same situation applies to many environmental problems. Pollutants that we can see, touch, and smell are disturbing enough, but we think that we can somehow handle them. Pollutants that none of our senses can respond to and yet can harm us are often more frightening.

Radioactivity falls into the latter category, and Figure 17 shows its most common manifestation. We've all heard of the rays that are emit-

134

FIGURE 17

A nuclear power plant tower on the Columbia River. This tower casts a shadow not only on the river but also on men's minds. The problem of radioactivity in our environment is a complicated one—perhaps the most complicated of any dealt with by this book. By using indices to evaluate each of the forms of radioactivity, we can make some sense out of the technicalities. (EPA—DOCUMERICA, Gene Daniels, 1972. Courtesy U.S. Environmental Protection Agency.)

ted from radioactive substances, but almost none of us have seen their effects personally. However, we've seen enough pictures of and statistics about victims of excess radioactivity to realize that these invisible rays can harm or even kill us. Interest in the indices of radioactivity has been stimulated by fear of the unknown. Not being able to see or touch an environmental pollutant doesn't make it worse—just a bit more mysterious.

One of the problems that we've had when dealing with other pollutants is how to measure their concentrations accurately. There's no such problem with radioactivity. Due to the efforts that have gone into nuclear physics in the past three decades, we can determine levels to a greater accuracy than practically any other aspect of environmental quality. For example, many radioactivity measurements can be made to tenths of a percent. We're often lucky if we can come within a factor of 2 to the correct count of wildlife populations.

There is a hitch, though. We can determine the concentration of this or that radioactive substance in the air or water easily enough. It's another matter to determine its effect on humans, and this is ultimately what we're concerned with. Because of this, standards for radioactivity effects on people haven't been as simple to devise as the corresponding physical measurements. As you'll recall, an index isn't really an index unless we have a standard or set of standards.

We recognize the symptoms of excessive radiation levels easily enough—burnt skin, leukemia, and even death. When the levels are low, the effects are less obvious, and problems arise in setting standards. For example, the average exposure of the U.S. population in 1970 to ionizing radiation was estimated by the Council on Environmental Quality to be about 182 millirems. (Units of radioactivity are discussed below.) This quantity was made up in part from natural sources such as radiation from the sun, and in part from man-made sources such as medical x-rays. Some of the man-made radiation—fallout from nuclear explosions and radioactivity produced from nuclear power stations— lowers environmental quality. However, these types of radiation produced a very low dose rate, as the effect on humans is called. The fallout effect was 4 millirems, and the effect of nuclear power was 0.003 millirems. While these numbers are small compared to the overall value of 182 millirems, they are not 0. Furthermore, the numbers are averages. Someone near the Nevada nuclear test sites is likely to be more concerned about fallout than a person on the coast of Maine. Living around the corner from a nuclear reactor will probably make you more conversant with millirems, neutrons, and all the panoply of nuclear fission terminology than if you lived hundreds of miles away. In addition to having more familiarity with the scientific jargon, persons near sources of radiation will probably have higher dose rates from this source than the national average.

The problem involved in setting radioactivity standards is that, although we see immediate results for high doses, the consequences of low doses aren't as immediately apparent. Many of the low dose effects are genetic in nature. Since one generation is about 20 or 30 years, a long time must elapse before results are clear. Another difficulty lies in the nonlinearity of effects. For example, a dose rate of X millirems per year for 5 years, or a total of $5X$ millirems, doesn't always produce the same result as a rate of $5X$ millirems for 1 year, the same total.

Since the units of radioactivity may seem strange, a few words about them can help. We can look at these units in two ways—from a physical or a medical point of view. To take an example from left field, we can compare the two cases to baseball. The physical units, which deal with the number and types of rays and radioactive particles that are emitted, can be compared to the number and types of pitches—curves, fastballs, knuckleballs—that the pitcher throws. The medical units, which depend on factors like the part of the body, the age, and the sex of the person exposed to the radiation, are analogous to the results of these pitches on the catcher. A fastball might affect his catching hand in a different way than a curve, and a wild pitch could even knock him unconscious. For example, one physical unit is a curie, which is a measure of the number of atomic disintegrations per second. A medical unit is the rem, an acronym for *radiation equivalent—man*. The rem is a rough average of the effects of different types of radioactive particles and rays on human organs.

A further complicating factor is the large number of radioactivity sources. We usually consider three types of rays, alpha, beta, and gamma, although other rays and particles such as x-rays and neutrons can affect us. Rays coming from different elements possess different energies and thus different effects on humans. Some of the sources of radioactivity have some notoriety in the public mind. For example, strontium-90, a substance produced in nuclear bomb tests, was on everybody's lips and in everybody's milk in the 1950s. There are dozens of these isotopes—forms of elements that have different atomic weights and radioactive properties from the "normal" elements we're accustomed to—each with its own peculiar effect on humans. No wonder that producing a single standard or a small set of standards for radioactivity has proved difficult.

As if the situation weren't complex enough, the age and sex of the person who is exposed to the radiation are important in setting standards. For example, people past the childbearing years are unlikely to be concerned about genetic effects. Standards must somehow average the effects on older, middle-aged, and young people.

In spite of having as many *caveats* as a Philadelphia lawyer's contract, it is possible to devise some simple indices. The Canadian environmental index contains radioactivity indices, divided into three sections, that measure its effects in air, whole milk, and water near nuclear reactors. Let's see how it was calculated.

The index of air radioactivity considered measurements of only one type of radioactivity—beta rays. Since there were no official standards for air, the measured levels were compared to the "background" levels. The background levels are those that would exist even if we didn't produce atomic bombs and nuclear reactors. The radiation comes from the sun, outer space, the rocks and bricks around us, and other uncontrollable factors. Most of the increase in radioactivity levels in air over background has been produced by atomic tests. The Canadian index for air radioactivity was quite low because at the time of measurement only the French and Chinese had the dubious privilege of watching mushroom clouds.

The index of radioactivity in milk also evaluated some effects of atomic tests. The two isotopes, or sources, measured were strontium-90 and cesium-137. This measurement was not an index, since the radioactive level of milk is ordinarily 0. As we have mentioned before, an index is essentially a fraction. If the denominator is equal to 0, the fraction has no meaning. The milk measurements could only compare one place to another, not how these places compared to a standard scale.

Since the signing of the Test Ban Treaty by some of the major nuclear powers in the early 1960s, the worldwide number of atomic tests has dropped drastically. We still need indices like those described above on milk and air, but they're a bit like locking the barn after the nuclear-powered horse has escaped. What concerns many of us is the level of radioactivity emitted from nuclear reactors. It seems likely that, for the rest of this century and beyond, we shall be increasingly dependent on this source of energy. We then require an index to tell us how the radioactivity that is leaking from them affects our environment.

The Canadian index considered only the waters around nuclear reactors. Because of their cooling requirements, the stations are usually situated near a river, lake, or ocean. As in the previous two indices, most of the radioactivity comes from natural background. Comparatively little is emitted from the reactor. In consequence, the index was also low, although as more stations are built, this may not always be so.

Radioactivity can permeate so much of our surroundings that we can consider it to be a pocket-sized edition of all environmental quality indices. For example, previous chapters have divided indices into those dealing with air, water, and land. We can do the same with indices of radioactivity, and the Canadian work omits only land radioactivity. However, land indices can easily be calculated. For example, the MITRE study reported that the Soil Conservation Service in the United States measures the strontium-90 levels of soil. In addition, foods such as milk are tested for radioactivity levels in many countries. An index based on this measurement would be comparable to the pesticides index discussed in Chapter 6, in that both attempt to determine relative levels of pollutants in food.

Radioactivity has been a matter of concern ever since Hiroshima. Although the number of bombs tested has decreased, this has been at least partially made up for by nuclear reactors and less publicized sources of radioactivity such as medical x-rays and color television. By using a number of indices of radioactivity, we can better understand the magnitude of the effect from all these sources.

YOU KNOW I CAN'T HEAR YOU—NOISE INDICES

Someone once suggested that if, as seems quite evident, modern civilization is equivalent to noisy civilization, it's only because of our piety. After all, doesn't the Bible admonish us to "make a joyful noise unto the Lord"?

The increased sound levels in our homes, streets, and places of work have been blamed on a wide variety of factors, but religion has not usually been one of them. Regardless of who's at fault, the loss of the sound of silence is one of our greatest misfortunes. Figure 18 shows one sufferer. Is it possible to devise an index for something as fleeting as noise?

FIGURE 18
What's just part of the job to the driller is excruciating noise to a passerby. At the first assault on our eardrums, noise is just noise. Yet we can analyze it in more detail to calculate noise indices. Since most of the sounds we encounter are not as rattling as this jackhammer, we need more accurate measures than how fast we clap our hands over our ears. (EPA—DOCUMERICA, Erik Calonius. Courtesy U.S. Environmental Protection Agency.)

Noise has been described as unwanted sound. This statement has more to it than the usual *bon mot*. We can measure sound levels quite accurately. However, noise has both a physical effect on our eardrums and a psychological effect on our brains. The notes A and B played together on a piano or the howling of a dog shouldn't bother us. Somewhere there may be some lost tribes to whom these sounds are delightful, but to most of us they're just noise. Our ears don't differentiate between Mozart and a kitten on the keys, but our brains do.

In many ways the calculation of an index of noise has many of the problems of that of an index of radioactivity. For example, the study of noise uses many units to describe human reaction to sound. The physical unit of sound has been standardized as the decibel, but noise has been measured in units of phons, sones, noys, and perceived noise levels, as well as a host of others. Trying to explain the definitions and advantages of each would take pages. Each in its own way tries to take some account of the psychological basis of noise perception.

Another similarity to radioactivity indices is the distinction between physical measurements and the effects on humans. For both noise and radioactivity the instruments that determine the physical levels are precise enough. Finding out how our bodies react is considerably more difficult. We don't have dials in the middle of our foreheads that indicate how much we are suffering from the two environmental hazards.

The pair of pollutants differ in that much noise pollution lasts for only a few seconds or less. This attitude wouldn't be shared by a 40-year-old parent enticed to a rock concert by a teenaged son or daughter, but it's true in many instances. For example, an airplane zooming just over your rooftop may leave your ears ringing, but the whole event is over in a triflingly short time. The jackhammers in the street may set your teeth on edge, but most of us, with the exception of the jackhammer operator, can get out of its range in a minute or so. The presence and effects of radioactivity usually persist for a much longer length of time. The fallout after a nuclear test can drift down from the sky for decades. Some sources of radioactivity can produce rays at almost the same rate for thousands of years.

The short span of much noise pollution is one of the reasons for the complicated units that are sometimes used. The units are mathematical ways of measuring long-term effects of short-term noise.

In spite of the efforts that have gone into noise studies, governments haven't conducted many surveys from which an index of noise can be calculated. The levels of noise vary so much over time and space that obtaining an accurate average is a problem. For example, a 1949 Studebaker that hasn't had a new muffler in years may drive down your street at 2 A.M. How do we average your fall out of bed with the quiet of the rest of the night, when you could easily distinguish between the alto and soprano sections of a cricket chorus? We might be able to give a tentative answer if we had enough instruments on enough street corners, but right now we don't.

As a consequence of this relative inaction, we have sets of standards for noise, but comparatively little data. The fraction that would be the index has a denominator, but not much of a numerator. This is in contrast to the situation for many other environmental indices we've discussed, for which there have been heaps of data but no official standards.

Let's look at some that have been set for noise. In Japan, a series of 21 different allowed levels were used. These were based on the proximity of an area to roads, the number of lanes in the road, the function of the area (business, residential, or hospital), and the time of the day or night. For example, the strictest (or lowest) standard was 35 phons for night hours in a medical area, and the highest standard, 65 phons, was for daytime in areas bordering on major roads. The phon is a unit of equal loudness for different sound frequencies.

Noise levels were also evaluated in the Manchester study. However, this work was based only on estimates, not measurements. The noise level in a street is usually roughly proportional to the concentration of cars moving on it. However, this estimate doesn't take account of that unforgettable 1949 Studebaker, whose roar drowns out everything else in the neighborhood. Noise estimates are fine until our ears are assaulted by old clunkers and low-flying aircraft.

In the calculation of an overall index of noise, the MITRE report tried to take into consideration such factors as the time of day or night, the type of activity, and the location of noise sources. The reasoning is too complicated to mention more than briefly. The method assumes that people with different activities are exposed to different levels of noise, and compensates mathematically for this. The jackhammer operator we referred to a few paragraphs back will never encounter the low noise levels that a university professor does. However, a decrease in the level of jackhammer noise produces an improved value of the street noise index even though the professor never hears the commotion from his ivory tower.

Once a crude index of noise involved comparing a source to that of a "public nuisance." For example, a yowling cat in someone's back yard was evaluated in a patrolman's mind against the noise he would allow in a reasonably quiet neighborhood. This method may have provided some mental gymnastics for police and judges, but the system was clearly inadequate. Now noise-control agencies throughout the world are using sophisticated equipment to match the noises that we hear against acceptable levels. All that we now need to devise indices of noise is to compile this mass of data into simple and usable forms. These eventual indices alone won't eliminate the sound barrage our ears experience all too often, but they will tell us when and where the worst clatter, clamor, and general pandemonium occur. We can then go to work with silencers and mufflers on the racketeers around us.

SHAKE, RATTLE, AND ROLL—VIBRATION INDICES

The jackhammer referred to in the previous section was more than noisy—it shook the teeth in our jaws. Unless we can devise indices of vibration as well as noise, we are only dealing with part of the problem.

Anyone who has ever passed by a subway or other rapid-transit facility in a large city realizes that there is more annoyance from trains than just the sound. On hot summer days people often mumble to each other, as if it would help, "It's not the heat, it's the humidity." By analogy, we can say, "It's not the sound, it's the vibration."

Although our cities sometimes shake so much that they seem about to fly apart, surprisingly little effort has gone into studying this problem. We know that vibration passes through the earth, and sound through the air. Some work has been done in California and other earthquake-prone areas to determine the effects of vibration on buildings, but trucks rumbling in the streets seem to have escaped the notice of most environmental scientists.

One of the most determined efforts to understand vibration is reported in *Tokyo Fights Pollution* (1971). As yet the Japanese have only standards and comparatively few measurements. How do we set a standard for vibration? Like setting noise standards, the problem is complex. We evaluate acceleration, or the change of velocity. When vibration occurs, the earth moves in a way not envisioned by Hemingway; it has a velocity. Even near a strong source of vibration, the ground can move only so far before the stationary part of the earth stops it. The moving ground then changes velocity—which is an acceleration. A large change in velocity means a large acceleration, which we feel as a strong sense of vibration. Small accelerations are usually associated with barely detectable vibrations. By using a standard based on acceleration as well as our psychological reaction to it, we can determine how bothersome vibration really is.

SOCIAL AND ENVIRONMENTAL INDICES

The environment is only part of our environment. This seemingly contradictory statement does have some meaning. Many of the indices we've considered have been traditional in that they deal with what we ordinarily think of as the environment—air, water, land, and wildlife. If we adopt a more general view and think of the environment as including all of our surroundings, we end up with a larger group of indices.

Let's use an example to illustrate the concept of these *quasienvironmental* indices. We've mentioned a wide variety of air pollutants that could be used to calculate indices. They've all been characterized by their being measured outdoors. This type of measurement would be adequate if we always lived outside, but in truth we spend much of our time inside offices and factories. Indoor air pollution isn't treated in the outdoor indices. Considering the state of the air where we work in turn leads us to the health effects of employment. An index of on-the-job health is what we can call a quasienvironmental index—it is only partly based on our usual ideas of the environment.

Goldsmiths claim that they can hammer that metal so thin that we can actually look through it. Making an ounce of gold cover a wide area may be an advantage, but stretching a logical point that thin can be a drawback. Environmental indices extend into more nooks and crannies of life than we've discussed in the past eight chapters. But, if we try to imply that they cover everything, we spread ourselves so thin that the holes in the argument become all too clear. The next two sections cover a few areas in which the relationship between environmental indices and other measures of our surroundings is strong.

ON-THE-JOB HEALTH

We are exposed to environmental threats more often than the number of times we walk down the street or go swimming. Most of us spend more than half of our hours indoors. The environmental problems can be worse inside.

Let's start with our house or apartment. We don't ordinarily think of the air inside as being polluted except perhaps after a session of cooking onions. The old saying about gravity is, "What goes up must come down." We can change it to read, "Some of what goes up in the air outside comes down inside." When the outdoor air index indicates poor conditions, part of the pollution enters our homes. Only a few air measurements have been taken inside houses, but they indicate that the problem can indeed be serious. Just as many of us have indoor-outdoor thermometers, we may eventually need indoor-outdoor air quality indices.

We don't usually produce many harmful air pollutants in our houses. This isn't the case for many factories and other places of work, where air pollution is all too common. For example, in the last few years alarm has been expressed about the industrial levels of vinyl chloride (VC) and asbestos fibers in the United States and Canada. Since these pollutants are generally not measured past the factory gate, we have no city-wide indices for them. If we had the chance to compute indices for VC in factories, values would be high in many cases.

In many places of work an analogous situation occurs with respect to noise levels. Boiler factories are not known to be good places to listen for dropping pins, but investigations have shown that noise in less publicized plants can be just as disturbing to sanity.

One reason that we don't have many environmental indices dealing with places of work is that, in the Western world at least, most factories are private. Managers usually aren't overjoyed at having inspectors come around with strange-looking gadgets to measure air quality and noise levels. Laws and attitudes have been gradually changing, and more enlightened employers have realized that measurement of environmental quality is in their own best interest as well as that of the workers.

How can we devise environmental indices for factories and offices? For noise the problem is easily solved. The United States has already set standards for allowable levels in places of work. The measurement of factory noise levels tends to be more precise than in the street, since the sound tends to be of the same type and relatively constant. One difficulty associated with indices of outdoor noise is that the sound levels vary from one second to the next in type and intensity. A pounding jackhammer is followed by silence that in turn is broken by a roaring engine. We can usually devise noise indices for factories much more simply than we can for cities.

The problems with air indices are considerably greater. Many of the factory air pollutants are unusual, and little is known about their long-term effects on health. For example, we have devised some simple standards for common pollutants like sulfur dioxide on the basis of many statistical and medical tests around the world, but who knows the results of breathing a few parts per billion of vinyl chloride? The bits and pieces of knowledge we have about some of these exotic pollutants are not sufficient for the many standards we need. Until we have these standards, we can't calculate the indices to tell us the state of a factory's working environment.

These few examples haven't exhausted the ways in which we can produce on-the-job environmental indices. Odor, vibrations, and a host of other factors in environmental quality can be measured and, in some cases, compared to standards. Our appreciation of our surroundings doesn't stop when we report for work.

BEYOND THE GNP—SOCIAL INDICATORS

Over the past few years, many a politician's speech has been larded with emphasis on the quality, as opposed to the quantity, of life. Exactly what is meant by these fulsome phrases is rarely made clear, but they may mean that we should concentrate less on the dollar and more on the way we live. A number of countries have already issued statistics to supplement the usual economic ones like gross national product and balance of payments, usually under the title of "social indicators." Environmental indices are obviously one type of social indicator, but there are many others. We can't discuss all of these indicators, but some are related to environmental fields.

As an example, take health statistics. Most sets of social indicators contain data on life expectancy, causes of death, the types of disability, and a host of other details. Environmental conditions affect many of these statistics. Excessive air pollution, for instance, may cause disease and even death. We can't yet determine what proportion of disease is caused by our carelessness in disposal of wastes in air, water, and land. In spite of this, we can view health statistics as quasienvironmental information.

We haven't said much about devising indices on the basis of health data. Good measurements are available, but the standards are still lacking. In terms of life expectancy, should we use the example of Methuselah or something more mundane? We may want an illness standard of no disease, but short of adopting the ancient practice of paying one's doctor only when we're not sick, most of us can't see how this could be done. It's clear that indices of health have yet to be perfected.

If the concept of environment is extended far enough, it enters into a wide variety of social indicators. For example, our homes are certainly part of our environment. Social indicators consider such factors as the structural soundness of housing, the presence of facilities like indoor plumbing, and the degree of overcrowding. The Canada land index mentioned in Chapter 6 uses an overcrowding subindex, and other housing indices can be calculated. For example, one index could be the number of dilapidated houses divided by the total number of dwellings. Some of these housing indices may not be environmental indices as we've previously defined them, but they do evaluate an important part of our surroundings.

The streets we walk on—or avoid—at night also form a portion of our environment. The United States publication on social indicators shows a wide variety of statistics on crime and other social conditions. Eventually we may need indices of social, as well as strictly environmental, wrongdoing. By this point we may have stretched the ounce of gold we referred to on p. 144 too thin to be useful. Still, it is better to have calculated a few indices that describe only part of the problem than never to have calculated at all.

The environment is more than the number of fish in a stream or the haziness of the air. In one sense, it is an attempt to describe our total surroundings—the way we live, work, and play. The further away we get conceptually from air, water, and land pollution, the thinner the ounce of gold that we stretch, but sometimes it's worthwhile to hammer away.

TWO CAN PLAY THAT GAME—INDICES AND IMPACT STATEMENTS

Here we enter the land of sheep, for all around us is wooliness. Environmental impact statements have received far more public notice than have indices. In spite of this, most environmentalists are hazy about how they work. Since impact statements deal with many of the same problems as indices do—a combative person might even suggest a contest between the two concepts—we should compare their strengths and weaknesses to those of indices.

What are impact statements? The Warner book, listed in the Bibliography, lists three main types: the Battelle approach, the matrix method, and the Water Resources Council (WRC) way. Although these differ in methodology, they all have as a main objective the assessment of environmental changes resulting from a specific project. The project could be a new road, building, or other public work such as a nuclear power station.

Indices are concerned with the state of the environment over a wide range of geography and time. Impact statements deal with one particular area and one particular period of time—before and after a project is completed. Because an index covers the whole of the United States doesn't mean that it can't specifically cover Houston, for example. In fact, just as we can't have a completed jigsaw puzzle without each piece fitting into place, we can't calculate a national index without first determining local indices. In this view, we can consider impact statements to be a category or type of index.

This seems to imply that anything that impact statements can do, indices can do better. Not always so; these statements make a much greater effort to predict than do indices. Elsewhere in this chapter, we note that indices have only rarely been used to tell what will happen in the future, although there's nothing to prevent them from being used this way. Since the main reason for impact statements is to foretell what is to come, they are clearly superior to indices in this respect.

We haven't really said what impact statements are. Let's consider the three types mentioned, without going into great detail.

The Battelle statement is based to a large extent on measurement. A total of 78 environmental parameters are considered, and a graph, or "value function" (shown in Figure 19), is drawn for each. The graph shows environmental quality ranging from 0 to 1 versus the parameter

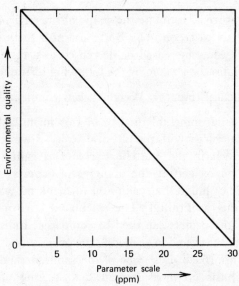

FIGURE 19
Illustration of a Battelle environmental impact statement. In this example, the parameter or physical scale is in parts per million. It could also be in acres of parkland, miles of visibility, units of water turbidity, or any of dozens of other units. If we know that the concentration of the pollutant being considered is 25 ppm, we can then simply read off the value of environmental quality from the graph. In this particular case, the relationship between environmental quality and the physical scale is linear—*but it isn't always this way.*

scale, which is in physical or chemical units. This may sound complicated, but it isn't as involved as it appears. For example, suppose that we were considering cadmium in drinking water. The physical scale would be the concentration in parts per million. Environmental quality would be high for a low concentration and low for a high concentration. (The scale used in the Battelle statement runs in the opposite direction from the ones we've discussed in this book, but that's only a minor mathematical point.)

The Battelle formulation does not make clear whether all of the 78 parameters, which encompass aspects of air, water, land, and other environmental variables, are considered for every project. For example, if we were considering construction near the ocean, it would be difficult to determine its effect on hoofed wildlife.

If we can measure or estimate the shape of the environmental quality graph, we're ready to go on. The Battelle aim is to find whether environmental quality has increased or decreased as a result of a proposed project. It calculates the change by the following formula:

Environmental Impact = Weight × Environmental Quality.

We find the environmental quality in this formula from the graph that shows how it varies with the physical scale, as we mentioned above. The weight, which in the Battelle parlance is called *parameter importance,* is found by getting the opinions of experts in the field. For example, they may judge that cadmium in drinking water is twice, or half, as dangerous as lithium. The calculated environmental impacts before and after the project can then be compared. Indication of significant loss is a signal to abandon or modify the project.

The Battelle approach uses some of the same methods as do indices: weighting, emphasis on measurement, and scaling of environmental quantities by graphs or equations. It can thus be viewed as a useful adjunct to these indices.

The matrix approach (shown in Figure 20) takes a different path. You can think of a matrix as a crossword puzzle without any black squares. Each of the blank boxes represents the interplay between an environmental action and environmental characteristic. For example, suppose that highway construction were being considered. The third action down the vertical axis of the matrix might be the draining of a swamp across the proposed highway's meanders. The second characteristic along the horizontal axis might be the aquatic mammals, such as muskrats, in the area. The draining of the swamp would ruin the muskrat colonies. This would be recorded as a numerically high value in the small box at the intersection of the third row and second column. If we were considering the effect of the swamp draining on the characteristic of lead in the air, we'd conclude that there was no effect at all. The appropriate box would probably be left blank to indicate this.

A total of 100 actions and 88 characteristics are available, leading to a total of 8800 possible interactions. It's beyond the mind of man to evaluate all of these, so only a fraction are used for each project.

The matrix approach suffers from the following problems: First, choosing a few interactions out of the 8800 can be bewildering. The example we gave of neglecting lead over a swamp is obvious enough, but most possible choices lie somewhere on the borderline. There probably

ACTIONS	Lead in air	Wild mammals	Outdoor recreation	Rare plants	Turbidity in water	Radioactivity	Noise			
Canal building		3		1	6					
Automobile traffic	10		2			8				
Swamp drainage		7	5	3	1					
Can recycling			2							
Pulp and paper mills			7	2	8	5				
Park construction		10	10	7	2					

(Column header: CHARACTERISTICS)

FIGURE 20

The matrix approach to environmental impact statements. We've only shown a small part of the whole matrix, since there are 100 possible actions (vertical) and 88 possible characteristics (horizontal). In sum, 8800 possible interactions, represented by boxes in the matrix, are then possible. The higher the value in the box, the stronger the interaction between the two factors. For example, it's unlikely that canal building would change levels of radioactivity anywhere, so the box at the intersection of this action and characteristic is left blank. On the other hand, there is a strong relationship between lead in the air and automobile traffic, since this action is virtually the only source of the pollutant. As a result, the value in the box is 10, or the maximum. One of the main problems of the matrix method lies in the wide divergence in the ways people evaluate the interactions, and fill in the boxes. For example, I have decided that the interaction between swamp drainage and rare plants should have a value of 3. A botanist in the next county, concerned about a unique mushroom, may give it a value of 10. We don't yet have a way of reconciling these differences.

isn't any one person or committee that could adequately determine which boxes should be filled.

Second, the exact number placed in a box can be based on measurement, the evaluator's judgment, or some combination of the two. No particular approach to finding the number is specified. As a result, another person filling out the boxes might arrive at different values using the same evidence. In the jargon of statisticians, there are problems of *replication*.

The main advantage of the matrix approach is that it "lets it all hang out." There are few environmental interactions that can't be found somewhere in the 8800 boxes. However, this method suffers from its lack of quantitative analysis.

The WRC impact statement is the simplest of the three. Environmental actions are classified into four broad areas dealing with such factors as open spaces and pollution. Impacts on these systems are evaluated in terms of size or other quantitative measures, descriptive or qualitative information, and possible improvements to the project. However, no instructions are given as to how the person doing the work is to judge between all the possible variables. The replicability of this method between one person and the next is even lower than that of the matrix method. The WRC approach can be summarized as "write down everything we know about the problem and hope that it makes some sense."

Scores of papers have been written on these and other types of impact statements. Their quality varies widely. Contrary to that hypothetical combative person mentioned in the first paragraph of this section, there is no inherent conflict between indices and impact statements. Because these statements try to evaluate all of the environment, known and unknown, they tend to be vaguer—woolier—than indices. Indices are more precise, but to get that precision some less-exact information must be skipped over.

Both indices and impact statements have a role to play in evaluating the environment. If we realize the limitations and advantages of each, we'll move farther along the road to effective action.

CONCLUSIONS

Environmental indices have aspects that include physical, psychological, and physiological effects. Radioactivity and noise are two of the more prominent examples of less obvious fields. In addition, by allowing the tight definitions of indices to loosen somewhat, we can see that they relate to a wide variety of social conditions. Our calculations should be more than a set of numbers; they should describe and help us understand a wide range of human activities.

10

WHERE DO WE GO FROM HERE?

Environmental indices, as described in the last seven chapters, are one of the best sets of tools we have to assess the state of the environment. But the general public must become as interested in the overall environment as in startling headlines that describe environmental disasters. These indices serve the same purpose as an editorial writer for a newspaper. The editorialist has to look beyond the big print to weigh the merits of a public issue. Numbers are sometimes not available on which to judge a political or social issue. But indices may yield as clear a picture of the environment as a wise and experienced editorialist provides on questions of the day.

HOW THE GENERAL PUBLIC FITS IN

Environmental indices can be a complicated and abstruse subject. To someone who has problems adding up his grocery bill, the mathematics used in some indices may be hair-raising. However, this doesn't mean that the public can't contribute to their formulation. In fact, if the public doesn't take part, indices won't be as effective as they could be.

It used to be said that war is too important to be left to generals. The state of the environment is too important to be left to scientists and administrators. There obviously has to be a scientific contribution; without it, our understanding would be perilously close to zero. But in constructing environmental indices there is room for the voice of a well-informed public.

Let's see how this voice can be heard. First, there has to be pressure to create environmental indices in the first place. As a scientist myself, I don't belittle the contribution of my fellow researchers. Too many of them, though, are not interested enough in communicating to the public the results of their laboratory experiments. As movie casting directors used to say, "Don't call us. We'll call you." One-way communication worked fine decades ago when environmental problems were comparatively insignificant, but not now.

Every year sees the publication of tens of thousands of reports, studies, and papers describing in one way or another the state of the environment. To a person without some scientific training, the work is usually almost unintelligible. More effort on the part of the authors would make much of it clearer to most of us—even to those who struggle in vain through high school chemistry. Producing environmental indices can be regarded as a way of simplifying and standardizing these mountains of often-unread information, so that the people who pay the piper can have at least one or two notes wafted occasionally in their direction.

It's one thing to say that the general public must act; it's another to say how. The most obvious way is through conservation and ecology groups like the Sierra Club in the United States, Pollution Probe in Canada, and Friends of the Earth in the United Kingdom. These groups have already done excellent work in alerting all of us to environmental hazards. Occasionally their outcries have sounded a little shrill to some, but this seems to be the only way to get attention in our society.

Now their role is changing. We need a watchdog to let us know that burglars of environmental quality are in the neighborhood. Unfortunately for this quality, there are far more burglars than there are watchdogs. Because of this, we also need a way of measuring how often and where the thieves have struck. We require these measures, or indices, not because we're satisfied with the rate of burglary, but so that we can deploy the watchdogs to nip a few bandits in the act.

There are already signs that some environmental groups are studying more objective measures of our surroundings than their press releases and demonstrations might indicate. For example, the National Wildlife Federation in the United States, mentioned in Chapter 3, has for some years issued simple environmental indices. Other organizations are also moving in the direction of cold logic as opposed to heated debate. With the threats to our environment as serious as they are, and human nature as inflammable as it is, there will always be the temptation to pounce on a menace to our ecology. Environmental indices will help us control the number and strengths of our pounces.

The first method of encouraging environmental indices could be summarized as making sure that conservation groups publicize them. Since these groups are often the only ones concerned with environmental issues, their influence is great.

The second method is more direct. Ordinary citizens can speak to their elected representatives, at the local, regional, or federal level about the state of the environment. A few of these representatives are as confused as some of us. But many of them are reasonably familiar with economic indices such as the gross national product and the consumer price index, since that's what they spend part of their time debating. It may not have occurred to them that analogous indices can be produced for the environment. Sometimes a gentle push on a seemingly immovable representative of the people can produce motion.

And what of the bureaucrats themselves? Because of the complicated nature of gathering and interpreting national environmental data, it's likely that governments will be the ones that calculate indices. As in any large bureaucracy, there is resistance to new ideas in most of the onion-like layers of government. It's always easier to pile up heaps of information than to commit yourself by explaining what it all means, which is what environmental indices imply. Sitting on the fence can eventually make one's bottom sore, and in pressuring for indices we are really doing government a favor.

The public doesn't often have the dubious honor of meeting members of the civil service. If you do, don't be bamboozled by a bureaucrat. Due to the nature of the system in which they work, they have been trained to give a thousand reasons why a new idea can't work and only one half-hearted one for why it just possibly could. Environmental indices are important enough for you to demand that the one overcome the thousand.

The role of the public can go much farther than merely pressure. We can all have a part in actually devising the indices. How can this be done? At first glance environmental indices may appear to be just tools used by scientists. While it's unlikely that the public can set standards, they can help to set weights.

The word *standard* has been sprinkled throughout this book like cherries in Mom's fruit cake. As you'll recall, the standard for an environmental quality is merely the denominator in the fraction we call an index. For example, a standard for an air pollutant might be X ppm, and for a water pollutant Y ppm. These standards are set on the basis of medical and health reasons, and detailed scientific reasoning is often used in their specification. No room for the layman here.

The setting of weights is a different matter. By *weights* we mean the relative importance of one subindex compared to another when they are mathematically combined. Chapter 2 talks extensively about the difference between weights and standards, and how it works in the consumer price index. The public, in effect, sets the weights of the price index by its choice in the market place. Our preferences, and index weights, can change. Until the advent of granola cereals in the early 1970s, wheat germ and rolled oats were associated with health freaks and the barnyard, rather than the breakfast table. Their weight in any food index would have been almost nonexistent. If their weights were now calculated by the statisticians, it's certain that they would be much larger.

Let's apply this reasoning to environmental indices. Right now, we have only crude methods for combining subindices to produce an overall index. For example, the Canadian environmental index was primarily divided into three parts, dealing with air, water, and land. There was no obvious way of assigning different weights to these three areas, so it was decided to assign an equal weight to each. This may seem simple-minded, but no other practical scheme presented itself.

We can use public opinion to obtain more representative weights. For example, the people of Japan have been extensively polled on which forms of pollution are the most important to them. In Tokyo prefecture, 30% said air, about 8% water, and about 17% odors. The other 45% had other concerns. An overall environmental index for Tokyo would have to take account of these "public" weights to at least some extent.

People's reactions are bound to differ depending on where they live and how they perceive their environment. In the same poll, people in another prefecture (Shizuoka) appparently thought that air pollution wasn't too bad. Only about 8% considered it important. To residents of this prefecture, water and odor had about the same importance that it did to residents of Tokyo. A Shizuoka environmental index would place less emphasis on air than a Tokyo index, but the other two components would be about the same. If we wanted to compute an index for all of Japan, we would have to take into account the feelings of all the different prefectures about the different sources of pollution, as well as their relative populations. This sounds like a formidable task, but the statisticians who compute economic indices have punched similar sets of numbers into their calculating machines for years.

To summarize, there are at least three ways in which the public can move environmental indices forward. We can work through conservation groups, encourage our representatives and civil servants, or participate directly in the setting of weights by expressing our concerns. All of these actions will hasten the day when indices provide us with better environmental understanding.

TRIAL AND ERROR, BUT NOT TOO MUCH OF THE LATTER

The old saying has it that change is the only thing that is constant. This will apply to environmental indices as they become used. For example, a scientist may discover that lead or carbon monoxide is more dangerous to health than had been previously thought. If the standard for one of these air pollutants had been X ppm, this finding might suggest that a new standard should be $0.7X$, $0.5X$, or in general some quantity less than X. Since the standard is in the denominator of the fraction that is the index, the index will increase (get worse) even though the measured values remain the same.

This seems like an obvious drawback to environmental indices. It is. But it doesn't destroy their value. It's true that this change in standard would make the environmental state seem worse. For example, if the standard went from X to $0.5X$ ppm, the index would double.

This problem is common to any index of human activity. To take account of changes in our tastes, the weights for the consumer price index are recomputed every decade or so. Recall the granola example of the previous section. The price index for 1947 isn't useless because the statisticians of the day used different weights. Comparing the present total price index to that of 1947 is difficult because of all the changes that have taken place, but we can still compare the cabbage subindex of a generation ago to the present one. A green head has remained a green head.

Standards and weights are bound to change every so often for environmental indices, as well. There is some consolation, though. First, many of these changes are presumably toward greater scientific accuracy in our environment. This property of more precision isn't shared by economic indices. For example, the standard for a pound of cabbage may have gone from 20¢ in 1951 to 25¢ in 1961. This doesn't indicate anything about the relationship of humans to cabbages, or vice versa, except that the price went up slightly in a decade. Similarly, the question used to poll the work force on the extent of unemployment may change, leading to higher or lower apparent rates than before. This alteration doesn't show that more or fewer workers are looking for jobs, but that people react differently to different questions. Changes in nonenvironmental indices often don't have as much significance as they do for environmental measures.

Second, we're usually interested in specific indices, not the overall value. For example, we may be interested in the index of sulfur dioxide for a particular city. If the standard for this pollutant is altered, we can easily recompute previous values to produce consistent results. Recalculating every component of an overall air quality index is a more massive job, especially if many standards have changed. Given enough computers, it can be done.

If we want to know how the overall environmental quality index for the United States or Sweden has varied over a long period of time, the inevitable changes in standards and weights, to say nothing of the addition of new subindices, will make the task a challenge. Comparing this year's particulate matter index for San Diego or Boston to last year's will be simple even if the standard has changed in the interval. Checking different time spans and degrees of combination falls between these two levels of difficulty.

As we build on our present rudimentary store of environmental knowledge, changes in standards and weights will become less frequent. Indices will undergo many transformations before they're fixed in, at best, a semifinal form. Nobody likes trial and error associated with a matter of as deep public concern as measuring and evaluating the state of our surroundings. If we are to succeed in our quest for accurate understanding, we must tolerate some barking up the wrong environmental trees.

I SEE A TALL, DARK, HANDSOME STRANGER—PREDICTION OF INDICES

Madame Lazonga's crystal ball may have foretold the inevitable tall, dark, and handsome stranger, but how are environmental indices predicted? These indices can be used to tell us the state of a part of, or even all of, the environment. Although we can determine what trends have held in the past, knowledge of future conditions seems to be unattainable from these indices.

If the past is highly forgettable and the present highly confused, the future can be regarded as highly unknowable. The simultaneous world crises in food, energy, and population seem to bode ill for the environment. But in a limited way we can use indices and other information to predict trends.

Let's see how this process could work. Suppose that the index for sulfur dioxide in the air for Minneapolis has decreased for a number of years. Since the levels of this pollutant are based to a large extent on the concentration of sulfur in oil, it doesn't take too much sophistication to deduce that a fair part of the index drop is probably due to a lowering of the sulfur levels in the oil being used. So far, we've made a rough correlation between the fall in the index and the likely cause—in the past.

Suppose now that the wells supplying the low-sulfur oil dry up, and to keep warm in the winter the city has to rely on oil containing large concentrations of sulfur. A bit of calculation shows that the sulfur dioxide index will rise. The amount of increase depends on the sulfur levels in pre- and post-dry well days. If the citizens of Minneapolis can see that an air index is going to rise, it may be worth their time, trouble and money to search for other sources of low-sulfur oil.

The future need not always look so bleak. As another example, consider an index of hydrocarbons in the air. Most of this pollutant comes from automobile exhausts. Suppose Minneapolis is offered a mass-transit plan that would decrease by 20% the average number of miles driven by car. This percentage can be used to make estimates of the contraction in the hydrocarbons index. It won't fall by the same proportion as the drop in automobile miles due to other factors, such as hydrocarbons produced by other gasoline-using machinery and changes in automobile engines. The predicted decrease in the hydrocarbon index gives the Minneapolis taxpayers a number with which to compare the cost of the mass-transit system.

Economists have been able to predict, with reasonable success, how such indices as the consumer price index change with world conditions. For example, when the price of crude oil rose dramatically in the early 1970s, the inflationary effect on the price index could be calculated to within a few percent. Not all changes in economic indices can be foretold, any more than all changes in their environmental counterparts can be. For example, even if we had perfect statistics on present conditions nobody could predict how much inflation we shall have in the next decade. Similarly, even if environmental indices were in full operation, they couldn't be used to predict all possible trends. If our lackadaisical attitude toward our wastes continues, the indices aren't likely to get much better. If we improve our attitudes and actions, there's a reasonable chance for index improvement.

Being able to foretell the future has been a prized attribute for thousands of years. In Biblical times, the most revered sages were known as prophets. Our more complicated society has made it increasingly difficult to tell what tomorrow will bring, let alone next year. By using environmental indices, we can lift the curtain slightly on some parts of the future, and make it serve the present.

GETTING YOUR ENVIRONMENTAL DOLLAR'S WORTH

Financial analysts going through a company's annual report don't spend too much time on the colorful pictures of smiling employees and sparkling factories. They usually head for the back page, where they find listed the annual expenses and income. If they're really in a rush, they'll skip even these figures and proceed to the bottom financial line— the profit or loss.

In the field of government, there's no concept analogous to this figure. True, governments do take in a certain amount in taxes and usually spend more than that, but this frequent budget deficit is not exactly a measure of what has been achieved in a year. Exactly what does the government do with our money?

Here's where environmental indices come in. They tell us clearly how much or little the environment has changed, regardless of how many dollar bills have been scattered over the landscape. For example, a government may claim that it has spent x million dollars to solve a particular problem in water pollution. In spite of inflation, most of us are rather impressed by that sum. We'd be somewhat less bowled over if the national or regional index of this aspect of water showed no change during that year. In effect the index, not the amount of money spent, would be the bottom line.

We've been using the word *government* as a short form for national, regional, and local governments, corporations, nonprofit organizations, and private individuals. Government is only part of the solution to environmental problems. The exhortation that all of us, in our homes and places of work, must attack these questions together has perhaps worn a bit thin by this time, but it's still true. As mentioned in previous chapters, governments are likely to be the ones to gather and produce environmental indices, so the word really means "all of us."

Using indices as the bottom line in a measure of government performance isn't a new concept. For example, consider the health indices briefly discussed in Chapter 9. Governments spend literally billions of dollars on health research and services. What do they have to show for it except a pile of bills marked "Paid in Full"? If we used as one index the length of life compared to what it was a generation ago, it would have changed very little in that time. Its improvement has been about 6% over the past 20 years in North America—hardly enough to write home about. It's true that diseases such as polio and diphtheria have virtually disappeared, and there could well be other measures of health research efficiency. The simple concept of length of lifespan provides one bottom line, but as we work with this concept, we must be careful to consider other such lines.

Governments have traditionally been wary about the final reckoning. A highly paid executive may prefer to talk about how much he earns, rather than what he accomplishes. Similarly, governments would prefer to tell how much they've done to make a problem go away, not whether it's still around. In spite of this reluctance some governments have moved ahead on environmental indices. The results may be a little embarrassing at times but, since it's our money, we should know what we're getting for it—press releases or action.

ENVIRONMENTAL MOTION FROM A NOTION

Nonmathematicians usually don't get much pleasure from rows of numbers. We're more interested in seeing forward motion on problems than in adding up columns, yet we often must do our arithmetic before the difficulties can be solved.

Calculating environmental indices may give employment to a few mathematicians. That's not our real concern. We want to live in a better environment. If we made a graph of understanding versus action, we'd find the two curves to be quite similar in the long run. One of the main reasons that there haven't been as many solutions to environmental problems as we like is this lack of understanding. Indices offer one of the best opportunities for extending to people in all walks of life real understanding of environmental conditions. When the people awake, the polluters will tremble.

BIBLIOGRAPHY

The major books and monographs referred to in the text are as follows: C. A. Bisselle et al. *Monitoring the Environment of the Nation,* (MITRE Corporation, McLean, Virgina, April 1971, National Technical Information Service numbers PB 205 990 and PB 205 989); Secretariat Permanent Pour l'Etude des Problèmes de l'Eau, *Inventaire du degré de pollution des eaux superficielles rivières et canaux* (La Documentation Française, Paris, 1973); C. M. Wood, N. Lee, J. A. Luker, and P. J. W. Saunders, *The Geography of Pollution: A Study of Greater Manchester* (Manchester University Press, Manchester, 1974); Environment Agency, Japan, *Quality of the Environment in Japan, 1972* (Japanese Government Publications Service Center, Tokyo, 1972); Environment Agency, Japan, *Quality of the Environment in Japan, 1973* (Environment Agency, Japan, 1973); Council on Environmental Quality, *Environmental Quality—The Third Annual Report of the Council on Environmental Quality* (Government Printing Office, Washington, 1972); and H. Inhaber, *A National Environmental Quality Index for Canada* (forthcoming). Subsequent editions of the U.S. and Japanese publications, which are issued every year, should be consulted.

In 1971 the Council on Environmental Quality in the United States, as part of a massive study of environmental indices, commissioned a number of studies of specific cases. These include the following: D. W. Jenkins, *Development of a Continuing Program to Provide Indicators and Indices of Wildlife and the National Environment* (Smithsonian Institution, Washington, 1972); *Land Use Indicators of Environmental Quality* (Earth Satellite Corp., Washington); J. Strickland and T. Blue, *Environmental Indicators for Pesticides* (Stanford Research Institute, Menlo Park, California, 1972); *National Assessment of Trends in Water Quality* (Enviro Control, Inc., Washington, D.C., 1973); and C.

164

A. Bisselle, S. H. Lubore, and R. P. Pikul, *National Environmental Indices: Air Quality and Outdoor Recreation* (MITRE Corp., Washington, 1972). The proceedings of the AAAS symposium were published as W. A. Thomas, Ed. *Indicators of Environmental Quality* (Plenum, New York, 1972). The National Wildlife index was first published in *National Wildlife,* Oct.–Nov. 1971. Some other useful publications include the following: R. S. Greeley, A. Johnson, W. D. Rowe, and J. Grace Truett, *Water Quality Indices* (MITRE Corp., Washington, 1972); *Quality of Life Indicators* (Environmental Studies Division, Environmental Protection Agency, Washington, D.C., 1972); and *Guidelines for Water Quality Objectives and Standards* (Inland Waters Branch, Canadian Department of the Environment, Ottawa, Canada, 1972, Technical Bulletin No. 67). Impact statements are considered in Maurice L. Warner and Daniel L. Bromley, *Environmental Impact Analyses: A Review of Three Methodologies* (University of Wisconsin Water Resources Center, Madison, 1974).

The number of scientific papers that deal with one aspect or another of environmental indices is large. Any list noted here would soon be obsolete. Many of the more significant ones are listed in H. Inhaber, "Environmental Quality: Outline for a National Index for Canada," *Science,* Vol. 186, November 29, 1974, pp. 798–805. Reference 8 deals with papers on air quality; 9 with water quality; 10 with noise; 11 with wildlife; 12 with pesticides and 13 with radiation. Two other papers dealing with the Canadian environmental index and including a bibliography are H. Inhaber, "A Set of Suggested Air Quality Indices for Canada," *Atmospheric Environment,* Vol. 9, 1975, pp. 353–364, and H. Inhaber, "An Approach to a Water Quality Index for Canada," *Water Research,* Vol. 9, 1975, pp. 821–833. Most of the books mentioned above, especially the volume edited by Thomas, make extensive reference to scientific papers.

When I went to a large library to gather information on economic indices, I measured the catalog entries not in terms of numbers, but rather by space that was occupied by the cards in the file. Cards describing books on this subject consumed about two inches of the file, so I can only mention the more significant entries. Perhaps the classic in the field is Irving Fisher, *The Making of Index Numbers: A Study of their Varieties, Tests and Reliability* (Houghton Mifflin, Boston and New York, 1927, 3rd ed.). The following are some of the more recent books on the subject: W. R. Crowe, *Index Numbers: Theory and Applications*

(MacDonald and Evans, London, 1965); R. F. Fowler, *Some Problems of Index Number Construction* (Her Majesty's Stationery Office, London, 1970); and W. F. Maunder, *Bibliography of Index Numbers* (Athlone Press, London, 1970, 2nd ed.). The techniques used have spread throughout the world. An example of this is P. Mouchez, *Les indices de prix; étude methodologique* (Editions Cujas, Paris, 1961). Most of the books mentioned contain a discussion of the history of the subject, though perhaps not all of its limitations. Finally, in case you think that all is sweetness and light among the economists, there is *The Index Fraud* (All India Trade Union Congress, New Delhi, 1963). Clearly not all of the details of economic indices have yet been settled.

INDEX

Accessibility, 84-85, 87, 90, 100-101, 132-133

Acid, pollutant, 78

Acidity of water, *see* pH

Action on environment, 13, 162-163

Advertising, 56, 122
  outdoor, 122-123

Aesthetics, 19, 36, 47, 81, 123, 131-132
  disagreement, 123
  and indices, 27-28, 82, 85, 121-133
  see also Critics

Agriculture, *see* Farmers and farming

Air, 4, 6, 8, 17, 31-32, 40, 48, 57, 60, 67, 70, 96, 102, 105, 107, 136, 138-139, 144, 147, 149-151
  changes in, 49
  Donora, Pennsylvania, 32
  indices, 9-11, 26, 33-34, 38, 40, 43, 65-66, 71-74, 83, 86, 89-90, 125-126, 146, 159
  indoors, 144-145
  London, 32
  particles in, 42, 49
  pollution, 18, 20-21, 23, 33, 35, 38, 40-41, 44, 46-49, 58, 60, 63, 65, 70, 105, 109, 114, 118-119, 144-145, 147, 157-158
  quality, 1, 4, 6, 41, 46-66, 73, 84, 123, 125, 165

Airports, 63, 66

Airsheds, 73

Alberta, Canada, 96

Aldrin, 96

Algae, 68, 82, 116, 131-132

American Association for the Advancement of Science (AAAS), 35-36, 165

American Primitive painting, 105

Ammonia, 82, 127-128

Amphibians, 113

Animals, 104-105, 107-108, 112-114, 118-120, 151
  aquatic, 114
  see also Wildlife; *and particular varities, such as* Buffalo *and* Wolf

Antelope, 113

Anti-pollution groups, *see* Conservation groups

Arctic, 119

Arithmetic, *see* Mathematics; Indices, calculation

Arizona, 41, 132

Asbestos, 7, 48, 55, 145

Asia, 93

Automobile, 5, 57, 61, 151
  density, 38, 60
  engines, 60, 126

exhausts, 59, 160

Background, 6-8, 32, 138-139
  changes, 8
Bacteria, 75, 109, 116
Bargains, and price indices, 24-25
Battelle Institute, 148-150
Beauty, see Aesthetics and Scenery
Benefit-cost ratio, 38
Bible, 40, 84, 105, 117, 139, 161
Biochemical oxygen demand (BOD),
  43, 71, 75, 77-79, 82
Biochemistry, 107
Biological diversity, 108-111
Biological indices, 74-77, 82, 103-120,
  131. See also Wildlife; and particu-
  lar types, such as Lichens
Biological organisms, 78-79
Biological sensitivity, 105-108
Biologists, 105, 112, 114, 119
Biomass, 119-120
Biotic index, 110
Birds, 84, 97, 104-105, 109, 111, 113
  migration, 108
Birmingham, Alabama, 47
Bisselle, C. A., 88, 164-165
Blue, T., 164
Bluegill (fish), 107
Boating, 68
BOD, see Biochemical oxygen demand
Bond price indices, 19
Boston, 127, 159
Botanists, 119, 151. See also
  Biologists
Boy Scouts, 37
Bread, gross national product and, 62
Breweries, 75
Bromley, Daniel L., 165
Bronchitis, 17, 32
Buffalo (animal), 111-113
Buffalo, New York, 40

Bureau of Sports Fisheries and Wildlife, U.S., 113-114
Burn, maximum allowable (forestry),
  90

Cabbages, 28, 158-159
Cadmium, 76, 79, 96, 149-150
Cakes and ale, 97
California, 17, 108, 124, 143
Camping, 85-87, 89. See also Recrea-
  tion
Canada, 40, 43, 45, 51, 61-62, 66, 71-
  72, 75-78, 83, 86, 88-94, 96-98,
  100, 115, 138-139, 145, 147, 155,
  157, 164-165
Canals, 38, 43, 77, 151
Canary, 105, 107
Capistrano, San Juan, 108-109
Capital investment indices, 4
Carbon monoxide, 35, 40, 46, 49, 59-
  61, 125, 158
Cesium-137, 138
Chemical oxygen demand (COD), 79-
  80
Chesapeake Bay, 106
China, 92, 138
Chlorides, 75
Chlorophyll, 116
Chrome, hexavalent, 79
Cities, 12, 38-40, 42, 49-50, 63, 74,
  93, 99, 101, 105, 122-123, 129,
  143, 145-146, 159-160
  and indices, 12, 41, 72-73
  parks, 85-87
  suburbs, 39, 93
Coal, 98-99
  mines, 105
Coasts, see Seashores
Cod, 115
Coefficient of haze (COH), 40-43, 59
Coliform bacteria, 79

Colorado, 73
Columbia River, 135
Computer, 103
Conflicting information, 3-4, 13, 33, 44
Conservation groups, 1, 3, 29, 44, 108, 114, 155-156, 158
Conservation of resources, 31
Consumer price indices, 4, 7, 20, 23, 26, 53-54, 156-157, 160
  averages, 28
  city, 12
  errors, 24
  history, 15-19
  supermarket, 14
  weights, 23-25, 158
Copper, 79, 96
Corn, 108
Corrosion, 65, 83
Council on Environmental Quality (CEQ), 37, 39, 42-44, 82, 93, 98, 136, 164
Courts, law, 69
Crime, 36, 147
Critics, aesthetic, 121-125, 131
Crowe, W. R., 165
Curie (unit), 137
Currency, see Money
Cyanides, 7, 79

Dallas, 71
Data, 4-6, 9, 11, 28, 43, 45, 51, 63, 100-101, 142-143, 155
  costs, 6
  economic, 21
da Vinci, Leonardo, 125
DDT, 84, 96-97, 107
Death rate, 32. See also Health effects
Debris, see Garbage
Deflation, 20. See also Inflation
Denver, Colorado, 93, 101

Department of Commerce, U.S., 19
Department of Environment, Canada, 95, 165
Department of Interior, U.S., 113
Deserts, 84, 115, 120
Detroit, 40
Dimethyl sulfide, 127
Disbenefits, 22
Disease, see Health effects
Dissolved alkalis, 43
Dissolved oxygen, 39, 43, 71, 77, 79
  lag, 71
Dissolved solids, 39
Distances from environmental quantities, 100-102. See also Accessibility
DNA (deoxyribonucleic acid), 120
DO, see Dissolved oxygen
Donora, Pennsylvania, 32, 47
Double-counting of indices, 61
Dow-Jones index, 19
Dust, 6, 8, 41-42, 48, 65, 107
  Bowl, 94-95
  color, 42
  see also Particulate matter

Eagles, 80
Earthquakes, 143
Earth Satellite Corp., 39, 98
Ecology and ecologists, 42, 85, 108, 112-113, 155
  effects, 19
  food cycle, 97
  gross national product equivalent, 23
  news, 33
Economic indices, 2, 7, 10-11, 14-30, 53, 156, 158-161, 165-166
  assumptions, 30
  averages, 28
  bias, 28-29

history, 15, 18
limitations, 29-30
quantity versus quality, 27-28
theory, 29
weighting, 23-27
Economists, 1, 14, 18-20, 22-24, 27,
    29-30, 62, 160, 166
  experience, 30
  sophistication, 29
Effluents, water, 70, 76, 78, 80
Element, 48
Emerson, Ralph Waldo, 33
Emissions, *see* Estimates of emissions
Endangered species, *see* Wildlife
Energy, 9-10, 160
  crisis, 98
Engen, T., 126
Enviro Control Inc., 39, 82, 164
Environmental impact statements, *see*
    Impact statements
Environmentalists, *see* Conservation
    groups
Environmental Protection Agency
    (EPA), 43, 68, 81, 91, 104, 106,
    124, 130, 135, 140, 165
Enzymes, 107
Epiphytes, 118
E. Q. Index (National Wildlife Federa-
    tion), 34-35
Erosion, 8, 44, 84, 94-96
Escherichia coli, 75
Estimates of emissions, 57-58, 60-62,
    65-66, 71-72, 74-76, 78, 127, 142
Europe, 77, 85, 93
  Western, 45
Eutrophication, 116-117
Everglades, Florida, 91
Experts, 11, 35-36, 43, 66, 80-82, 119,
    150
Extinction, 111, 114
Extreme value index, 64-65

Factories, *see* Industry
Farmers and farming, 8, 21, 65, 69,
    84, 96, 108. *See also* Rural areas
Filthium, oxides of, 4-6
Fish, 11, 36, 47, 68-70, 76-77, 80, 84,
    97, 103-107, 109, 113, 115, 117,
    147
  kills, 37, 83, 117
  mercury, 44, 72, 115
  processing, 76
Fisher, Irving, 165
Fleetwood, Bishop, 15-17
Florida, 104
Fluorine, 119
Food, *see* particular types such as Milk
Food and Agricultural Organization,
    97
Food prices, 2-3, 10, 24-25. *See also*
    Consumer price indices
Food Prices Review Board, Canada, 14
Foreign trade indices, 19
Forests, 11, 34, 84-85, 89-92, 94, 100,
    118-119, 132
  fires, 90-91
  growth, 89-90
  indices, 42, 44, 89-92
  insect damage, 91-92
  redwood, 91
  reseeding, 42, 89-90
Fowler, R. F., 166
France, 43, 77-78, 83, 138, 164
  water quality, 43, 74-75
Friends of the Earth, 155

Garbage, 38, 68, 99-100, 129-130. *See
    also* Litter
Gas, natural, 98
  as pollutant, 48, 63, 105
Gasoline mileage, 60
Geologists and geology, 115, 131
GNP, *see* Gross national product

Gold, 20, 144, 147

Gourmetia, 2-3, 10, 24

Governments, 4, 7, 14, 37, 39-45, 52,
54, 62-64, 69, 75, 109-110, 114,
141, 156, 161-162

  bureaucracy, 156, 158

  and economic indices, 18-19, 29

  embarrassment, 44-45

  performance, 162

  spending, 161

Grand Canyon, 132

Grandma Moses, 124

Grand Teton National Park, 100, 132

Greeley, R. S., 165

Grit, see Smoke

Gross national product, 13, 19-22, 62,
146, 156

  as cumulative index, 23

  definition, 20

  unpaid services, 21

Gypsum, 74

Haddock, 115

Hal (computer), 103

Haze, see Coefficient of haze

Health effects, 6, 19-20, 31-33, 49, 51-
55, 59-60, 63, 67, 69, 105, 108,
118-119, 128, 136-137, 144, 146-
147, 157-158, 162

  and life expectancy, 146-147, 162

  and standards, 6-7

Heat pollution, see Thermal pollution

Hemingway, Ernest, 143

Henry VI (of England), 15-16

Heptachlor, 97

Herbicides, see Pesticides

Hours worked, indices of, 19

Housing, 147

Houston, Texas, 148

Humidex index, 2

Hydrocarbons, 17, 40, 49, 60,

126, 160

Hydrogen sulfide, 127

Hydrology, 131

Ice cream standard, 60

Idaho, 131-133

Impact statements, 148-152, 165

Impressionists, 125

Income, 17, 19, 21

Indices, calculation, 2, 5, 15-17, 24,
28, 31, 49-53, 62, 86

  changes, 34, 82, 158-159

  combination, 10-11, 18, 38, 40, 43,
56, 77, 157, 159

  cumulative, 23

  Gourmetia, 2-3

  historical, 53

  homogenized, 29

  how they work, 1, 165-166

  hypothetical, 37, 64-65

  and impact statements, 148

  imperfect, 44

  origin, 33

  prediction, 159-161

  primitive, 33

  revision, 45

  scales, 53-54

  unknown factors, 17-18

  unofficial, 19, 44

  users, 36

Industrial Revolution, 31

Industry, 4, 22, 49, 52, 62, 64, 76, 79-
80, 89, 99, 119, 126-127, 162

  areas, 6, 142

  development, 41

  financial analysis, 161

  and gross national product, 21

  output, 19

  pollutants, 44, 58, 61, 69-70, 73, 76,
78, 96, 128-129, 145-146

  profit or loss, 161

Inflation, 2-3, 12, 14-16, 20, 160-161
Inhaber, H., 164-165
Insecticides, *see* Pesticides
Insects, 107, 110, 113-114
Instruments, measuring, *see* Measurements
International indices, 12-13, 44, 58
International Union for the Conservation of Nature, 114
Isotopes, 137

Jackson units, 81
Japan, 44, 52, 59, 78-80, 83, 127, 143, 157, 164
    land, 86, 96
    noise, 142
    water, 78-80, 82
Jenkins, D. W., 164
Johnson, A., 165

Kansas, 101, 108
Kansas City, Missouri, 93
Kronecker delta, 50-51, 64-65

Lakes, 58, 69, 79, 103, 105, 117, 139
    age, 116
Land, 11, 27, 32, 37, 46, 58, 84, 94, 96, 98-99, 139, 144, 147, 149, 157, 164
    indices, 39, 43-44, 84-102
    and parks, 85
    pollution, 36, 38, 100
    productivity, 119
    quality, 86
    subsidence, 83
Landscape, *see* Scenery
Lead, 17, 48, 57, 60, 79, 125, 150-151, 158
    index, 57
    poisoning, 57
Leaves, tree, 110, 118

Lee, N., 164
Leopard, 113
Leopold, Luna, 85-86, 131-132
Lichens, 109, 118-119
Lilienthal, M. J., 88
Lindane, 97
Lithium, 76, 150
Litter, 7, 17, 37, 129-133. *See also* Garbage
Little Salmon River, Idaho, 132
Living space, 34, 36. *See also* Parks
London, England, 32, 47
Los Angeles, 17, 60, 73, 93, 119, 126
Lowe, Joseph, 17
Lubore, S. H., 165
Luker, J. A., 164

Maine, 136
Mammals, *see* Animals
Manchester, United Kingdom, 38-39, 60, 65, 164
    land, 99-100
    noise, 142
    water, 77-78
Maps, 72-75, 101
Marine euglene, 79
Marshes, *see* Swamps
Mass transit, 143, 160
Mathematicians, 14, 24, 59, 152, 157-158. *See also* Experts and Scientists
Mathematics, 34-35, 61-66, 76, 80-81, 84, 96, 98, 100-102, 105, 110, 113-114, 118, 132-135, 141-142, 146, 149, 152, 154, 162
    correlation, 18
    see also particular types, such as Root mean square *and* Kronecker delta
Matrix, impact, 150-152
Maunder, W. F., 166

Measurement, 5-6, 31-32, 36, 40-41,
    51, 57-58, 61-62, 64, 66, 69, 71-
    72, 74, 78, 82, 86, 93-94, 96, 98,
    103, 105, 107, 109-111, 114, 120,
    123, 125-127, 129, 132, 136, 141-
    143, 145, 147, 150, 152, 155, 158
    biological, 29, 92, 103
    chemical, 29, 75, 77-78, 81, 84, 92,
    103
    economic, 27, 36
    international, 12
    physical, 27, 29, 75, 78, 81, 84, 92,
    103, 141
    see also Data
Meat, 23-26
Mercaptans, 127
Mercury, 44, 72, 76, 115
Metals, in land, 96
    in water, 44
    see also particular types, such as
    Cadmium and Mercury
Mexico City, 126
Miami, 71
Michelangelo, 125
Michigan, 68
Microorganisms, see Algae; Bacteria
Milk, radioactivity in, 137-139
Minerals, 34
Minneapolis, Minnesota, 160
Minnow, fathead, 107
Mississippi-Missouri basin, 73
Mite, 113
MITRE Corporation, 36-39, 43, 48,
    50-51, 64-66, 82-83, 112-113, 116,
    129, 139, 142, 164-165
Moiwa, Japan, 79
Mollusks, 113, 115
Money, as dollars, 19-21
    lost due to pollution, 37
    and price, 27
    spent on pollution control, 4, 37,

161-162
    value of, 15
Monitoring, 4, 29, 69, 164. See also
    Data and Measurement
Monoculture, 108
Montana, 101
Montreal, Canada, 118-119
Mosses, 118-119
Mouchez, P., 166
Mount Fuji, Japan, 62
Mount Whitney, California, 86
Municipal sewage plants, see Sewage
    plants
Muskrats, 150

National indices, 1, 12, 13, 18, 35, 38,
    42-45, 58, 61-62, 66, 72, 75-77,
    82, 85-87, 114, 148, 161
    caution, 41
National Sanitation Foundation, 80-81
National Wildlife Federation, 34, 65,
    155
    E. Q. Index, 34, 165
Nebraska, 101
Nevada, 136
New Orleans, 73
Newspapers, 10, 14, 27, 33-35, 121,
    125, 129
    editorial, 154
    transitory nature, 33
New York, 40, 93, 126, 130
Niihama, Japan, 80
Nitrogen, organic, 82
Nitrogen oxides, see Oxides of nitro-
    gen
Noise, 36, 139-143, 145-146, 151, 153
    jackhammer, 140-142, 146
    pollution, 39
    psychological, 141
    as public nuisance, 143
    units, 141-142

North America, 77, 93, 98, 111-112, 122, 162
Nuclear energy, *see* Power plants, nuclear
Nuclear reactors, *see* Power plants, nuclear

Oak Ridge, Tennessee, 65
Oceans, 130, 139, 149
Odor, 36-37, 47, 63-64, 125-129, 133, 146, 157
  home, 38
  testing, 126
  see also particular types, such as Ammonia
Oil, 98, 117, 160
  burning, 49
  industry, 76, 117
  motor, 27
  slick, 81, 83
Ontario, Canada, Air Quality Index, 33, 40-41, 57, 59
Organization for Economic Cooperation and Development, 45
Osaka-Kobe, Japan, 80
Ottawa, Canada, 14
Overcrowding, *see* Population, density
Oxidants, 40, 49, 60
Oxides of filthium, 4-6
Oxides of nitrogen, 11, 17, 40, 59
  dioxide, 40, 82
  water, 82
Oxygen, 48, 67, 71

Panda, giant, 113
Parathion, 107
Paris, 74
Parks, 1, 8, 23, 44, 55, 84-87, 94, 100-102, 129, 149, 151-152
  indices, 85-86, 89
  national, 86, 88

public, 85
types, 84-85
visits, 88
Particulate matter, 40-41, 48-49, 56, 58, 63, 66, 159
  size, 49
  standards, 59
  water, 75, 79, 82
PDI index, 82
Perfume, 126-127
Pesticides, 39, 43, 92, 96-97, 107, 113, 139. *See also particular types such as* DDT
pH, 79
Phenols, 79
Philadelphia, 35
Phon (unit of loudness), 142
Phosphorus, total, 82
Pikul, R. P., 88, 165
Planners, 36, 41, 87
Plants, 63, 65, 97, 104-105, 107-109, 118-120, 151
  land, 36, 131
  water, 36, 70, 115-116
Politicians and politics, 12, 14, 40, 44, 52-53, 146. *See also* Governments
Pollock, Jackson, 125
Pollution and pollutants, 3, 6, 8-9, 17, 20-22, 26, 31-32, 51-54, 57-58, 60-61, 63-65, 67-69, 72-74, 76-79, 82-83, 90, 104, 106, 108-109, 112, 116-118, 120, 126-127, 134, 136, 141, 146, 152, 157, 159, 160
  changes, 49
  control devices, 50, 61
  distribution, 58
  double-counting, 62
  duration, 82
  exotic, 146
  and indices, 8-9
  levels, 33, 44, 49, 64, 69

per capita, 75
synergy, 56
see also particular types, such as
   Thermal pollution. *In addition,*
   see under media of pollution, such
   as Air; Water; *and* Land
Pollution Probe, 155
Polychlorinated biphenyls (PCBs), 80
Population, 37, 160
   density, 92-94, 101, 147
Porter (19th century economist), 17
Power lines, 133
Power plants, 5, 70
   nuclear, 9, 135-136, 138-139, 148
Prices, 20
   history, 15-19
   see also Consumer price indices
Proximity, *see* Accessibility
Public, 62, 69, 75
   access to land, 84
   action, 155-156, 162
   confidence, 52
   consciousness, 31, 33, 48, 154, 159
   contributions, 154
   information, 39
   opinion, 157-158
   relations, 4
   and scales, 55
   warning, 41
Pulp and paper, 76, 126, 151

Quality of life, *see* Social indicators
   and indices

Radioactivity, 9, 36-37, 43, 134-139,
   141, 151, 153
   dose rate, 136-137
   units, 137
Rays, radioactive, 136-138, 141
   alpha, beta and gamma, 137
Recreation, 1, 9, 67, 76, 83, 87-90,

147, 151, 165
Recycling, 151
Red Sea, 117
Red Tide, 78-80
Redwoods, 91-118
Refuse, *see* Garbage; Litter
Regions, 40-41, 75, 78, 85-88, 99, 115
   differences, 42
   and indices, 1, 33, 38-39, 42, 61,
      161
   and standards, 65
   and weights, 26
Regulating agencies, *see* Governments
Rem (radiation equivalent-man), 136-
   137
Reptiles, 113
Research, 4, 7, 9, 18, 43, 52
Riggins, Idaho, 133
Rivers, 38, 47, 52, 58, 69-72, 74, 77-
   80, 82, 85, 94, 103, 107, 109, 116,
   131-132, 139, 147
   basins, 72-75, 80
   Canadian, 76
   flows, 73
   French, 43
   healing action, 72
   tributaries, 72
Riverside, California, 93
Roads, 129, 142, 148, 150
Rodents, 113
Root mean square (mathematical tech-
   nique), 51, 65-66
Rowe, W. D., 165
Rural areas, 62, 73

Sailing, *see* Boating
Salmonella, 75
Salmon River, Idaho, 133
San Diego, California, 159
San Francisco, 40
Saunders, P. J. W., 164

Scales, 33-34, 38, 40-41, 53-55, 63,
    77, 122-123, 128, 150
  bidirectional, 55
  inverted, 63
  linear, 65, 149
Scenery, 84-85, 87-88, 131-133. *See
    also* Aesthetics
Scientists, 4, 17, 34-36, 39, 44, 48, 52-
    55, 108, 118-119, 133, 143, 154,
    156, 158
  communication, 155
  disagreement, 124
  specialization, 36
Seashores, 69, 79-80, 84, 88, 115
Sedimentation, 94-96
Seto Inland Sea, Japan, 78-80
Sewage plants, 71, 76, 78
  pools, 72
  types, 72
Shellfish, 80, 82-83
Shizuoka prefecture, Japan, 157
Shrew, 113
Sierra Club, 155
Sludge, 69
Smell, *see* Odor
Smithsonian Institution, 39, 164
Smog, 17, 40, 60, 119
Smoke, 31, 38, 41, 48, 57, 65, 107
Smokestacks, 47, 127-128
Social indicators and indices, 36, 53,
    92, 144, 146-147, 153, 165
Soil, 34, 36, 74, 84, 94-97, 102, 119.
    See also Land
Soil Conservation Service, 95, 139
Solar radiation, 65
Solid wastes, *see* Garbage and Litter
Sound pollution, *see* Noise
Spending on environment, *see* Money
Spruce budworm, 91
Standards, 5-7, 9, 11, 16-17, 20, 42,
    44, 46, 49-51, 53-54, 59-60, 63-65,

    69, 74, 76, 79-80, 82, 86, 88, 90-
    91, 93-94, 96-98, 107-110, 114,
    116-117, 120, 125-128, 131, 133,
    136-138, 142-144, 146-147, 156-
    159, 165
  changes, 49, 51-53, 158-159
  of gross national product, 21
  international, 12, 45, 97
  of living, 15
  monetary, 10-11, 19-20
  regional, 65
  short-term, 64
  as yardstick, 19
Stanford Research Institute, 39, 164
Statisticians, *see* Mathematicians
Statistics, *see* Mathematics
Statistics Canada, 14
Steel, 61-62, 75, 80
Stock market, 2, 4, 10, 19-20
Streams, *see* Rivers
Streptococci, fecal, 75
Strickland, J., 164
Strip mines, 98-99
Strontium-90, 137-139
Subway, *see* Mass transit
Sulfates, 74
Sulfur dioxide, 5, 9, 11, 17, 35, 38,
    40-41, 44, 46, 51, 53, 56, 58-59,
    62, 65-66, 122, 146, 159-160
  standards, 59
Sulfur oxides, 59
Suspended solids, *see* Particulate mat-
    ter
Swallows, 108-109
Swamps, 116, 150-151
Sweden, 126, 159
Swimming, 67-68, 82, 130, 145
Synergy, 55-57
  defined, 56

Tama River, Japan, 110

Taxes, 21, 160-161
Temperature, 6, 69-70, 82
  air, 71
Thermal pollution, 23
Thomas, W. A., 35, 65-66, 80, 88, 107, 126, 165
Tiger, 100, 113
Timber, *see* Forests
Times Square, New York City, 101
Tokaichi River, Japan, 79
Tokyo, 46, 62, 110, 143, 157
Tosabori River, Japan, 79
Toxicity, 76
Trails, 85, 87-88
Trees, *see* Forests
Trimethylamine, 127
Truett, J. Grace, 165
Truffles and champagne, 2-3, 10, 24, 28

Unemployment, 4, 53, 159
United Kingdom, 77, 83, 99, 155
United Nations, 12, 97
  Environmental Program, 45
United States, 18-19, 37, 40, 42-43, 45, 59, 73, 113-114, 136, 139, 145, 147-148, 155, 159, 164
  land, 98
  water, 80-83
University of Manchester, 38
Urban areas, *see* Cities

Value function, impact statement, 148
Vibration, 143-144, 146
Vider (American scientist), 62
Vinyl chloride, 145-146
Visibility, 62-66, 149
  airport, 44, 62-63
Visual indices, *see* Aesthetics

Warhol, Andy, 125

Warner, M. L., 148, 165
Warning index, 64
Wastes, *see* Garbage; Pollution
Water, 17, 32, 46, 48, 57-58, 84, 96, 102, 107, 136, 138-139, 144, 147, 149
  appearance, 55
  dilution, 69-70
  drinking, 44, 67, 76, 83, 149-150
  hardness, 75
  and indices, 1, 9-11, 26, 34, 43-44, 58, 68-70, 72-74, 78, 81, 86, 89-90, 92, 115, 131, 165
  metals, 76, 79
  nutrients, 39
  parameters, 70-71, 74, 79, 82
  pollution, 6, 21, 26, 28, 58, 74, 82, 103, 118-119, 157, 161
  quality, 31, 39, 43, 67-84, 123, 131, 164
  table, 83
  turbidity and cloudiness, 39, 44, 76, 81, 116, 131, 149, 151
  uses, 67-69, 79
Water Resources Council, 148, 152
Weather, 37, 58, 69
Weighting of indices, 10, 12, 16, 23-27, 35, 41, 65-66, 72, 75, 77, 133, 150, 156-159
  changes, 26, 158-159
Whale, 113
Wheat, 108
Whooping crane, 111, 113
Wichita, Kansas, 101
Wildlife, 6, 8-9, 23, 34, 89-90, 104-105, 111-115, 120, 136, 144, 149, 151, 164. *See also particular types, such as* Birds; Buffalo
Williamsburg, Virginia, 122
Wolf, 111-113
Wood, C. M., 164

World Health Organization, 97
World Wildlife Fund, 114
Wyoming, 132

X-rays, 136-137, 139

Yellowstone National Park, 85, 100-
    101
Yosemite National Park, 88

Zinc, 96